The Information Warfighter Exercise Wargame

Rulebook

Christopher Paul

Ben Connable

Jonathan Welch

Nate Rosenblatt

Jim McNeive

Prepared for the Marine Corps Information Operations Center
Approved for public release; distribution unlimited
August 2021

NATIONAL DEFENSE RESEARCH INSTITUTE

For more information on this publication, visit **www.rand.org/t/TLA495-1**.

About RAND

The RAND Corporation is a research organization that develops solutions to public policy challenges to help make communities throughout the world safer and more secure, healthier and more prosperous. RAND is nonprofit, nonpartisan, and committed to the public interest. To learn more about RAND, visit www.rand.org.

Research Integrity

Our mission to help improve policy and decisionmaking through research and analysis is enabled through our core values of quality and objectivity and our unwavering commitment to the highest level of integrity and ethical behavior. To help ensure our research and analysis are rigorous, objective, and nonpartisan, we subject our research publications to a robust and exacting quality-assurance process; avoid both the appearance and reality of financial and other conflicts of interest through staff training, project screening, and a policy of mandatory disclosure; and pursue transparency in our research engagements through our commitment to the open publication of our research findings and recommendations, disclosure of the source of funding of published research, and policies to ensure intellectual independence. For more information, visit www.rand.org/about/principles.

RAND's publications do not necessarily reflect the opinions of its research clients and sponsors.

Published by the RAND Corporation, Santa Monica, Calif.
© 2021 RAND Corporation
RAND® is a registered trademark.

Library of Congress Cataloging-in-Publication Data is available for this publication.
ISBN: 978-1-9774-0723-8

Cover: U.S. Marine Corps photo by Lance Cpl. Kaleb Martin.

CONTENTS

FIGURES AND TABLES

Figures

Tables

1. INTRODUCTION TO THE INFORMATION WARFIGHTER EXERCISE WARGAME

The Marine Corps Information Operations Center (MCIOC) conducts an Information Warfighter Exercise (IWX) one to two times per year. Many recent IWXs (some under the prior exercise nomenclature *Combined Unit Exercise*, or CUX) have included a wargame-like "opposed free play" element in which teams of participants prepare plans for operations in the information environment (OIE) and then revise them based on friction encountered within a notional scenario and perhaps as opposed by other teams. In prior iterations, these wargame-like exercise elements have lacked a structured adjudication mechanism, which both trainers and participants have found unsatisfying. MCIOC asked RAND to help in developing a more structured wargame for IWX with a formal adjudication process. This document contains the ruleset developed, playtested, and implemented during the 2020 IWX cycle.

The IWX wargame is an opposed event in which two teams of players compete against each other in and through the information environment to better support their respective sides in a notional scenario. The rules for this wargame are scenario-agnostic and can be used with any appropriate scenario at any level of classification. See Section 3.4 for requirements related to the scenario.

Teams represent an Information Operations Working Group (IOWG) or information-related operational planning team (OPT), or its adversary force equivalent, as dictated by the scenario. Each team will have responsibility for generating a plan for OIE to support the objectives of the force of which they are notionally a part. The larger plan for the operation, including the overall commander's intent and scheme of maneuver, is largely fixed by the scenario and is part of the planning context.

Teams of players should bring a well-developed concept of support to the wargame from the prior week of IWX training (the wargame is intended to take place over three days, during the second week of a two-week IWX). Playtesting has revealed that an academic period (for cross-leveling and instruction) and a planning period prior to the start of the

wargame are essential to success. As part of the wargame, players will be called on to add details to their plan, amend that plan dynamically in response to in-game events, prepare discrete game actions as part of plan execution, and make cogent arguments in favor of their team's action and against the actions of the opposed team.

> OPTIONAL: Instead, the wargame might take place during a single-week IWX, with the academics and planning period considerably foreshortened or conducted partially before arriving at IWX, including read-ahead and perhaps remote presentations or discussion.

1.1. Using This Rulebook and Associated Materials

Sections in these rules are numbered for easy reference and cross-reference. The table of contents is a useful reference for finding rules regarding specific topics. Section 1 is the introduction and discusses the goals of the game, the terms used throughout the ruleset, and the overall flow of the game from turn to turn. Section 2 details each of the five steps of a turn within the wargame and is the most extensive section within the ruleset. Section 3 provides guidance for preparing to conduct the wargame—including staffing and roles required to conduct the game, details about scenario preparation (the wargame rules are scenario agnostic, but a scenario is required), preparing the physical spaces, game boards or maps, and representations or tokens for the game—and estimates of the time required for each step within each turn. Section 4 summarizes a few additional considerations related to the conduct of the game.

1.1.1. Optional Rules

Throughout this document, some rules—such as the one higher up in this column, about a one-week IWX—are flagged as OPTIONAL and set in blue boxes. Any of these rules may be included in any

IWX wargame at the option of the trainers running the game. As a practical matter, the lead of the group of exercise controllers will have the final say regarding which optional rules are in play and which are not.

1.1.2. Downloadable Player's Guide

This rulebook is intended for those running the IWX wargame. A brief *Player's Guide* is available at www.rand.org/t/TLA495-1, and players should be given that document rather than this full rulebook (lest better understanding of the nuts and bolts of the game lead them to "game the game"). Players can be sufficiently prepared to play through verbal instruction and the *Player's Guide*.

1.1.3. Downloadable Game Aids

Various scoresheets, checklists, and other printable game aids (GAs) are available at www.rand.org/t/TLA495-1. In this rulebook, we refer to these by number and name—for example, *GA1: Turn Structure Overview* and *GA3: Judge's Responsibilities*. Table 1 lists the name and number of each game aid.

1.2. IWX Wargame Learning Objectives

1.2.1. IWX Overall Goal

The overall goal of IWX is to improve participants' understanding of how to plan and execute selected elements of Marine Corps OIE. There are seven functions in Marine Corps OIE (listed in the annex at the end of this rulebook), and IWX places primary emphasis on "influence foreign target audiences," with secondary emphasis on "deceive foreign target audiences," "inform domestic and international audiences," and sometimes, depending on the mix of participants, "attack and exploit networks, systems, and information." It might be possible to include other functions of Marine Corps OIE within this game structure, but doing so might require some improvisation or adjustments to the rules. **IWX (and this wargame) is intended to focus on efforts to influence, inform, or deceive in support of operations.**

TABLE 1

Downloadable Game Aids

#	Name
GA1	Turn Structure Overview
GA2	EXCON Lead's "Run of Show" Summary
GA3	Judge's Responsibilities
GA4	S2 Role-Player's Responsibilities
GA5	S3 Role-Player's Responsibilities
GA6	S3 Role-Player's Action Approval Checklist
GA7	Narrator's Preparation Worksheet
GA8	Engagement Scoresheet
GA9	Head-to-Head Engagement Scoresheet
GA10	Target Number Calculation Record Sheet
GA11	Action Results Record Sheet
GA12	Making an Outcome Roll

1.2.2. IWX Secondary Goals

Secondary goals of IWX include a desire for participants to continue to develop OIE-related planning skills that will aid them when they return to their home units and when they participate in real OPTs in the future. Specifically, a goal of IWX is that participants will improve their presentation and briefing skills and their ability to articulate and defend their concept of support.

1.2.3. Objectives of the IWX Wargame

This wargame supports the goals described in Sections 1.2.1 and 1.2.2. It does so by allowing play in an environment where IWX participants execute their concept of support within the confines of a realistic scenario and against an opposed team that is seeking to execute its own concept of support. Both sides will be called on to adjust their plan to conform to realistic obstacles and in response to the actions of their counterparts or to otherwise react to situations that arise within the game and scenario that may be outside of the scope of their original plan.

During the game, success, failure, and mixed results will all be possible. The effectiveness of player planning and actions will be determined by the quality of the plans prepared by each side; their respective ability to present actions based on those plans; friction between opposing plans; friction between actions and the scenario environment; expert adjudication; and chance. In order to succeed, and as part of the game, teams of players must

1. Clearly describe their target, target audience (TA), or audience and the desired effect on that audience
2. Continually reassess the information environment and contend with unavoidable ambiguity in the information environment
3. Articulate how selected information-related capabilities (IRCs) are to be deployed and offer reasons for the effectiveness of their employment
4. Explain how proposed actions affect the target, TA, or audience and how that effect supports their side's scheme of maneuver and overall mission success
5. Verbally present their moves in a clear, concise, and well-thought-out statement with appropriate supporting justifications
6. Defend the argument for the success of their actions against the counterarguments of their opponent, if required
7. Present verbal counterarguments to their opponents' presented actions.

1.3. Terms

This section presents and defines the terms used throughout the ruleset. Terms are listed here and used consistently throughout to avoid confusion. Terms are not presented alphabetically but rather are aggregated by topic area (people in the game, structure and sequence, presentation and adjudication) and sorted within topic area based on primacy.

1.3.1. People in the Game

Participant: Someone attending IWX as part of the training audience.

Player: A participant at IWX who is playing the wargame. Players and participants should entirely overlap unless a participant is excluded from the wargame for some reason (such as only being present for a portion of IWX).

Side: The side in the conflict in the wargame scenario for which the teams play.

Team: The group of players representing an OIE OPT or equivalent for one of the two sides in the wargame scenario. There are two sides and therefore two teams. Ideally, each team should have no more than 10 players and no fewer than 6. Each team represents an OIE OPT or IOWG and *only* influences the plans and actions of their side in the conflict so far as they realistically would be able to as an OIE OPT.

BLUE: The side that is playing the U.S. Joint Force, Marine Corps force, or allied or partner nation force in the scenario.

RED: The side that is opposing BLUE objectives and pursuing its own objectives. For this wargame, players for the RED side will still use U.S. Marine Corps and/or joint planning doctrine and templates to produce their plans and actions.

Exercise Control (EXCON): The adjudicators, judges, and scenario managers responsible for maintaining all elements of the game environment that are not directly under the control of the players. EXCON will be drawn from MCIOC personnel and other subject-matter experts (SMEs) as available and required. EXCON controls the maneuver elements of both BLUE and RED and all HQ elements (at all echelons) other than the OPTs represented by the teams. EXCON also controls any other groups or forces in the scenario context that might otherwise be a separate exercise cell (GREEN, for example).

WHITE: In wargaming circles, EXCON is sometimes referred to as "WHITE."

GREEN: Relevant noncombatant individuals and other groups and actors within the scenario. GREEN is not a separate team and is controlled by EXCON. EXCON may include an individual specifically responsible for planning and tracking GREEN actions or response, or this may be a general EXCON task.

Role-player: A member of EXCON who portrays a member of the staff of the RED or BLUE side outside of the teams. The wargame requires, for example, a role-player to play the operations officer (S3), to whom the teams brief their intended actions during Step 3 of each turn. Other role-players, such as an intelligence officer (S2), could be required, depending on the scenario.

Narrator: A member of EXCON who describes (narrates) both (1) the outcomes of actions based on adjudicated outcome rolls and (2) the overall progress of the operation at the end of each turn. The narrator tells the story of the game as it unfolds from turn to turn. Some of that story is pre-scripted storyline within the scenario and based on maneuver plans, and some of that story is dynamically determined by the actions taken by the teams and the success or failure of those actions.

Target audience (TA): Throughout this ruleset, the subject of an action or proposed action is referred to as the *target audience*. However, there are certain circumstances where it would be inappropriate to refer to the subject of an action in that way; for example, when considering actions to defend the friendly force or the American public from adversary disinformation or communications disruption. In such situations, players are urged to use appropriate nomenclature (public, audience, etc.), and EXCON members are encouraged to mentally make appropriate substitutions within the rules (considering an audience or protected formation where a scoresheet calls for a "TA," for example).

1.3.2. Structure and Sequence

Turn: A cycle through the sequence of steps in the game. A turn includes receipt of an update about the scenario environment and progress of the operation; preparation and approval of multiple actions; engagement and discussion of all of those actions; and adjudication of the actions. The IWX wargame typically consists of six turns, and each turn should take between 3.5 and 6 hours to fully complete. Turns represent periods of time within the game scenario, but the amount of time represented by a turn may vary throughout the game. For example, turns when the tempo of maneuver forces is relatively low may represent longer periods of time.

Step: Each turn consists of five steps: (1) receive scenario and situation update, (2) prepare to present, (3) present actions for approval, (4) engagement and matrix debate, and (5) results and reset. Steps 1–3 are performed in spaces specific to each team, whereas Steps 4–5 take place in a shared engagement space.

Action: One of the building blocks of a turn. Within each turn, each team will prepare and present a number of actions. Each action must have a target, TA, or audience and a desired effect. An action can include multiple IRCs in combination. Actions are argued for and against by the two teams during the

engagement step. Actions are named and referred to by EXCON based on the team that presents them and the order in which they are presented, so action RED 2 would be the second action presented by RED in a given turn. The desired and allowed number of actions per team per turn can vary at the discretion of EXCON, but will usually be either two or three and should be announced by EXCON during Step 1 of each turn.

Continuing action/recurring action: An action that is begun in one turn and then continues on into later turns. Continuing actions are recognized as such by EXCON and may be subject to special adjudication. Continuing actions still need to be presented for (continuing) approval and presented and argued for and against in the engagement step.

Scenario: The notional but realistic context and operations that have brought the two sides into competition or conflict and provide the stage and field of operations for the wargame.

Storyline: How the overall scenario is expected to unfold except for the OIE parts. Essentially, the scenario will follow a preset script unless the teams' actions have a significant impact on progress toward one or more of the operational objectives. Even with OIE impact, the storyline is expected to remain within certain left and right limits.

Phase: Phases within a given game are defined by the scenario and the overall scheme of maneuver for each side that each team represents. Phases might include shaping, seizing the initiative, and decisive action. Individual phases in the operation depicted in the scenario will be pursued over one or more turns.

Inject: A change in the scenario introduced by EXCON during the game. Injects may be used to surprise or challenge the players, or to restore competitive balance if one team is doing exceedingly well or exceeding poorly early on in the game. In some games, they may not be used at all.

1.3.3. Materials and Setup

Game board: A display that represents the operational environment. Depending on the details of the execution of the game, the game board may be an actual game board, a plotter-printed map, or an electronic map. Whatever its format, the game board will include representations of the line of advance or progress of the maneuver aspects of the scenario operation, and it will provide opportunities for the two teams to geolocate their IRC forces and their targets, TAs, and audiences. During Steps 1–3, each competing team will work from a separate board in the team's own planning space. EXCON will maintain a board that contains ground truth for the position and disposition of all forces. EXCON will also prepare an engagement board to be present in the engagement room during Steps 4 and 5 that will depict some hybrid of RED and BLUE situational awareness and that will be used for matrix debate and adjudication.

Game aid (GA): Various scoresheets, checklists, and other printable game materials are available for download at www.rand.org/t/TLA495-1, and in this rulebook we refer to them by number and name—for example, *GA1: Turn Structure Overview* and *GA3: Judge's Responsibilities.*

Planning space: A room or other designated space in which team members can plan without the other team overhearing. RED and BLUE will each have their own planning space.

Engagement space: A room or other designated space in which the RED and BLUE teams meet for engagement and matrix debate (Step 4).

Control space: A room or other designated space for EXCON to operate in.

1.3.4. Presentation and Adjudication

Matrix system/matrix adjudication: A gaming system used to determine the success or failure of a given action. In a matrix wargame, adjudication centers on presentation and two-sided debate rather than on dice rolling. Dice rolls are used to help determine outcomes, but the effectiveness of each team's planning, presentation, and debate with the opposing team are the focus of the wargame and are important determinants of outcomes.

Presentation: Description of an action to begin adjudication, whereby the presenting player shares task, purpose, method (including product mock-ups), end state, and three reasons the action will succeed.

Presenting player: The player who is presenting an action. Players presenting for a team must be rotated, and the presenting player is the only one allowed to present arguments related to the action being discussed.

Rebuttal/rebut: Following the presentation of an action and a short discussion period, a single player from the non-presenting team offers up to three reasons that the action will *not* succeed. This is a presentation of reasons, *not* a counteraction. Rebuttal reasons may not include things that the non-presenting team would like to do to counter the action *unless* their approved actions or plans include a battle drill that would be triggered by the action.

Counterarguments: After the rebuttal, the presenting player may offer counterarguments. Counterarguments are limited in scope to being responses to the rebuttals from the non-presenting team. New points or issues may not be raised, only disagreements with or refutations of the rebuttal arguments are permitted.

Dice: Cubes with numbers from 1 to 6 represented on each face. In this wargame, three dice are rolled and added together. This will produce a result between 3 and 18 on each roll of the dice.

Outcome roll: For any adjudication event that has any element of chance, dice will help determine the outcome. The presenting player will roll three dice, add the results (for a total from 3 to 18), add or subtract any roll modifiers determined during the presentation and matrix debate step, and compare that with the target number set to accomplish the task effectively. This determines the outcome of the action or event. See Figure 1.

Target number: The number that a team is trying to reach or beat with an outcome roll. Target numbers are determined by EXCON and are a product of some of the events in the engagement step. If the total on the three dice plus the modifiers is equal to or greater than the target number, then the action succeeds. If the total on the dice plus modifiers is less than the target number, the action fails. The distance between the target number and the actual roll (plus modifiers) determines the degree of success or failure.

Roll modifier/bonus/penalty: Roll modifiers are either added to an outcome dice roll (a roll bonus) or subtracted from an outcome roll (a penalty). Roll modifiers may result from events in the scenario, the partial success of an action in a previous turn, or the outcome of the arguments for and against the action in the engagement step.

Degree of success/failure: Some actions or events might have binary results: outright success or outright failure. Other actions—to be determined by

EXCON—will result in varying degrees of success or failure. For these nonbinary events, the distance between the outcome roll and the target number indicates the degree of success or failure. For example, if the outcome roll (including modifiers) is 14 and the target number is 11, then the degree of success is +3 (14 − 11 = 3), or "success by 3." These numerical values for degree of success will have corresponding effects in the scenario environment and the overall adjudication and storyline produced by EXCON.

Reroll: Some optional rules allow for failed outcome rolls to be rerolled under certain circumstances. This is what it sounds like: a mulligan, a do-over. Some caveats apply: (1) When a team chooses to use a reroll, the new result is final; if it is worse than the initial roll, too bad. (2) When using a reroll, all three dice are rerolled (not just one or a subset). (3) Rerolls can only be used on rolls made by the team using a reroll—that is, you cannot use your reroll to force the other team to reroll their outcome roll. (4) Rerolls may not be used by either side in a head-to-head action (see Section 2.4.3.8.4).

> **NOTE:** The term *reroll* is not meant to apply to dice that fall strangely, such as when one falls on the floor or one leans against another die or other object and does not lay flat when the dice stop moving. When dice are thrown improperly (landing out of bounds or unsettled), they should not be read and they should not be thrown again. This does not count as a reroll, as the first roll was never properly completed.

FIGURE 1
Making an Outcome Roll

We roll three six-sided dice in this game to determine the success of an action. The dice are meant to simulate chance and factors beyond the control of the players and their role-played command staff.

Outcome Roll Procedure:

To make an outcome roll, use the following procedure:

1. EXCON announces a *target number*—the number that must be equaled or exceeded by the outcome roll in order for the activity to succeed. Target numbers will vary based on the difficulty of the action, quality of planning, and the matrix discussion.
2. EXCON announces any bonuses or penalties to be added to or subtracted from the roll, based on circumstances or player efforts.
3. Roll the dice!
4. Add up the three dice, then add bonuses or subtract penalties, to get your outcome roll total.
5. If your total roll meets or exceeds the target number, you have succeeded! If it is less than the target number, your action has failed. (Depending on the specific game rules laid out by EXCON, you might have the opportunity to reroll.)
6. EXCON determines the degree of your success or failure based on the difference between the target number and your outcome roll total.

Example:

During engagement (Step 4 in the wargame) a player from BLUE presents an action. After hearing the presentation and matrix discussion, the EXCON judges determine that the target number for the outcome roll is 12. The judges also note a special circumstance not covered in their criteria and announce that BLUE receives a roll penalty of −1.

The presenting player rolls the three dice and gets:

The dice add up to 15, from which 1 is subtracted because of the penalty. The outcome roll total is 14.

The outcome roll total (14) is greater than the target number (12), so the action succeeds!

The degree of success of this outcome roll is "success by 2" because the roll total exceeded the target number by 2. EXCON will determine whether success by 2 constitutes a "partial success," a "full success," or an "astounding success," depending on the action type and circumstances.

Tip:

When rolling and adding three dice, you have about a 91% chance of rolling 7 or higher, a 50% chance of rolling an 11 or higher, and about a 16% chance of rolling 14 or higher.

1.4. Overview of Play

This section describes the sequence of play and provides a general outline of the conduct of the various steps of a game turn. These steps are also summarized in Figure 2 and in *GA1: Turn Structure Overview*. Section 2 provides much greater detail on each step.

The standard IWX version of this wargame consists of six turns (other instances may involve a different number of turns, but the steps composing each turn would remain unchanged regardless of the total number of turns). Each turn should take between 3.5 and 6 hours to complete and is scheduled to occupy either the morning or evening of an exercise day. Time to complete a turn varies based on how quickly game play proceeds and based on how many actions per turn each team is allowed to undertake. See Section 3.6 for a discussion of estimates of time required for each step within a turn. During each turn, both sides execute all steps. Turns are simultaneous. Teams alternate in presenting actions during Step 4, engagement.

Each turn consists of five steps:

1. Receive scenario and situation update
2. Prepare to present
3. Present actions for approval
4. Engagement and matrix debate
5. Results and reset.

Each step is briefly described in the following sections and described in much greater detail later in this ruleset. Responsibilities of the EXCON judges across these five steps are summarized in *GA3: Judge's Responsibilities*, and *GA2: EXCON Lead's "Run of Show" Summary* provides a high-level summary of all EXCON and player activities across all five steps of a turn.

1.4.1. Step 1: Receive Scenario and Situation Update (Overview)

Step 1 takes place in each team's planning space. During this step, teams receive information on how many actions will be allowed in the turn (typically two or three); an update on the current state of the operation; how (if at all) it is deviating from the planned scheme of maneuver for their side; the cur-

rent locations of available IRCs; the status of GREEN; and any updates on injects or changing scenario conditions. The Step 1 overview summarizes the current state and lays out the intended destination for the turn for each side.

1.4.2. Step 2: Prepare to Present (Overview)

Still in their planning spaces, teams revisit their plans as needed and prepare their actions for the turn. For each action a team wants to take, it prepares a presentation that will be given during Step 3 (present actions for approval) to a member of EXCON who will role-play the team's operations officer (S3). This presentation should include arguments in support of the action, which the team will also use during Step 4 (engagement). Teams prepare mock-ups or working samples of any products or scripts involved in the execution of the proposed actions. The presenting player for each action rehearses their presentation and makes sure they are ready to offer compelling explanations and arguments, both to the S3 during Step 3 and to EXCON in general during Step 4.

1.4.3. Step 3: Present Actions for Approval (Overview)

In their planning spaces, for each proposed action, the presenting player (and only the presenting player) briefs the S3 role-player on the audience and intended effects, capabilities involved, and details of (planned) execution. A different presenting player presents each candidate action. The S3 role-player approves or disapproves each action. (If any of the approved actions require capabilities or permissions not organic to the team's side's forces, EXCON will immediately adjudicate whether the action can proceed using a special outcome roll.)

At the end of Step 3, the S3 role-player will leave the team's planning space to back-brief EXCON judges on the approved actions, so that the judges can review them and begin completing action scoresheets in preparation for Step 4.

FIGURE 2
Turn Structure Overview

Player Activities	The IWX Wargame: Turn Structure Overview		EXCON Activities
Step 1. Receive scenario and situation update			
Listen and take notes	Teams are updated on: » Number of actions allowed this turn » Map changes » Maneuver plans for the turn » Status of GREEN » Positions of available IRCs » Injects		Set game boards and number of actions allowed this turn S2s brief teams
Step 2. Prepare to present			

Plan/choose actions
Identify order of presentation
Pick presenters
Prepare presentations
Develop battle drills
	Each presentation should include:
	» TA/audience » Product mock-ups
	» Desired effect » Desired end state
	» Purpose/connection to » Assessment plan
	  commander's intent » 3 reasons the action
	» Conducting capabilities will succeed
			Observe team preparations
			S3s prepare to rule on
			  Step 3 proposals
			Complete RFI matrix as needed

Step 3. Present actions for approval			
Present actions to S3s for approval Finalize presentations of approved actions Prepare for rebuttal	» S3 approves or disapproves proposed actions » Roll for higher-level approval (if needed) » EXCON preps for Step 4		S3s review and approve/disapprove actions S3s and observers brief judges Judges prepare draft scorecards
Step 4. Engagement and matrix debate			
Present actions Rebut other team's actions with 3 reasons why they will fail Counterargue rebuttals Make outcome rolls	For each action: » Presenting team presents *(5 mins)* » Other team prepares, rebuts *(4+2 mins)* » Presenter counterargues *(2+1 mins)* » EXCON questions, scores » Outcome roll and result	For head-to-head actions: » First team presents *(5 mins)* » Second team prepares rebuttal *(4 mins)* » Second teams presents and rebuts *(6 mins)* » First team rebuts and counterargues *(4 + 2 mins)* » Second team counterargues *(2 + 1 mins)* » EXCON questions, assigns modifiers » Roll-off and result	Set game board Reorient situation Set order of presentation Finalize action scores Set target numbers Give roll modifiers Narrate results
Step 5. Results and reset			
Take notes Plot to sustain success or seek revenge	» Summary of all results » Share operation storyline to date		Summarize turn results Update storyline Adjust/prepare for the next turn

Planning Spaces (vertical label, rows 1–3)
Engagement Space (vertical label, rows 4–5)

1.4.4. Step 4: Engagement and Matrix Debate (Overview)

For this step, all players from both teams come together in the engagement space. Teams alternate presenting actions approved by their S3 role-player in Step 3. The same presenting player who presented actions in Step 3 presents them in Step 4. During engagement, each presenting player also presents three reasons why they believe their action will be successful. The opposed team then has a few minutes to discuss and presents three reasons why they believe the action will be unsuccessful or less effective than the presenting player has indicated. The presenting team will then have two minutes to pre-pare and an additional minute to present up to three counterarguments to the rebuttal. (This is a form of what is called *matrix adjudication* in the wargaming community.) EXCON will then have an opportunity to ask clarifying questions. EXCON will then quickly confer and finalize scoresheets, giving the presenting player a target number and any roll modifiers accumulated. The presenting player will then roll dice for an outcome roll, which will produce a degree of success or failure for the action. The narrator will describe what happens as a result of the action.

Teams alternate presenting actions, engaging in matrix discussion, receiving EXCON input and target numbers, and completing outcome rolls until all actions have been adjudicated.

1.4.5. Step 5: Results and Reset (Overview)

EXCON completes Step 5, recording all outcomes, narrating the results and consequences of all effects and determining impact on the progress of the overall operation. EXCON makes any adjustments necessary to the overall storyline and to materials that will be presented as part of the update in Step 1 of the next turn.

Steps 1–5 are repeated for all six game turns. After the final step of the final turn, EXCON announces the final outcome of the scenario operation, announces which team "won" (and why), presents any awards, and leads an after action-review to cement lessons learned and to seek input to improve the wargame (and IWX) in the next iteration.

1.5. Inputs and Activities Required Prior to the First Turn

Details on preparing this wargame, including scenario preparation and the selection of EXCON personnel to run the game, are provided in Section 3 of this rulebook. But we present the rules of the game first, in Section 2, and in that section we assume that the following things have happened as part of IWX prior to the wargame taking place:

- Participants have been divided into two teams of 6–10 players, one to represent the OIE OPT for BLUE and the other to represent the equivalent for RED.
- EXCON members have been chosen, mastered the rules of the game, and prepared a scenario. There is some flexibility in how many EXCON members there need to be, depending on factors such as the scenario and whether any EXCON members are serving multiple roles. Typically, EXCON will consist of S2 and S3 role-players for each team; three judges, one of whom is the "EXCON lead"; the narrator; a note-taker assigned to each team; and individuals with responsibility for taking notes, providing IT and network support, managing the game board, and serving as timekeeper.
- Both teams have received sufficient information to produce a concept of support, including relevant information about the context in which

the scenario operation will take place and the planned operations themselves.

- Both teams have been given a list of capabilities that are considered organic to their forces and that they can employ as part of their concept of support.
- Each team has been provided with a planning space—separate physical or virtual spaces each capable of accommodating the entire 6–10 members of each team plus 2–3 members of EXCON. Ideally, these spaces will include projection and network capability so that EXCON can display slides and players can collaborate, store material on a shared drive, and share and discuss slides and other content. It is also preferable that each of the two spaces also be the workspace for the respective teams.
- An engagement space (physical or virtual) has been prepared. This should be a space in which all players from both teams and all of EXCON can come together. It should not be either team's planning space or workspace, to avoid "home field advantage." The engagement space should have projection capability and network access so that presenting players can use slides if they wish.
- Game boards (either physical or digital or both) and appropriate tokens, markers, or other symbology have been made available to represent the current state of the scenario as understood by each side, as well as intended progress (both for maneuver and for OIE). As envisioned, the game requires four boards: one for each team planning space, one shared board for depicting actions and operational progress in the engagement space, and a ground truth/master board maintained in the EXCON control space to allow EXCON to track events.
- Dice will need to be available, with a surface or box on or into which they can be rolled (or a virtual substitute; there are a number of dice-roller web apps available).
- Relevant forms, templates, and scoresheets have been printed in sufficient numbers to support both players and EXCON for all turns.
- A timer is required. Ideally, this would be a large countdown timer that displays time

remaining across a large audience, so that all players and presenters can see how much time is remaining in a segment. Failing that, a stop-watch, kitchen timer, or other substitute can be sufficient but will require the timekeeper to give regular verbal warnings about time remaining.

- Should the wargame be scheduled to be played remotely rather than with all players and EXCON members together at the same site, additional materials and preparations may be required.

1.6. Example of Play from a Player's Perspective

This is a relatively complex ruleset with lots of moving parts. While the overview of play in Section 1.4 is a fine summary, it does not provide much of the texture of the game and may leave a first-time reader of this ruleset with a lot to take in. To make these rules more approachable, here we provide an example of play, narrated by a notional player:

Nineteen of us arrived at IWX last week and started receiving a bunch of good blocks of instruction on OIE. Midweek they told us that the whole second week would be a wargame and that we'd be developing an OIE concept of support for our side in the wargame. They broke us into teams; I'm on the BLUE team, which means that I'm one of 10 marines playing as part of the OIE OPT for a Marine Expeditionary Unit (MEU). The other 9 participants in our IWX class are playing as the RED team, which means they're like an OIE OPT for the Centralian 17th Mechanized Infantry Battalion. The mission of the Combined Joint Task Force—of which the MEU is a part—is to land in Montanya (a U.S. partner nation recently invaded by Centralia), secure an airfield, a port, and the nearby city to protect the welfare of the Montanyan population, roll back Centralian forces, and enable follow-on operations.

We spent the last few days of the first week going over all the intelligence we had on Montanya and the Centralian forces, familiarizing ourselves with the MEU's planned scheme of maneuver, and working on our OIE concept of support. We finished up our plans on Monday, gave a confirmation brief, and started the wargame on Tuesday. I guess it is true what they say about no plan surviving contact with the enemy, because things have not been going our way. Right away, turn 1 included an inject that one of our forward companies had gotten pinned down at an intersection,

lost several of their Light Armored Vehicles, and taken heavy casualties. That forced changes in the planned scheme of maneuver (but all of that was taken care of by EXCON), but also forced changes in how we were supporting that scheme of maneuver. We scrambled and adjusted our planned actions to re-target locations where our marines actually were (rather than where they were supposed to be if they had advanced on schedule), and came up with a clever feint and supporting information efforts to relieve some of the pressure on the marines fighting at the intersection.

Now, it is Wednesday morning of the second week of IWX, and we've just started turn 3. We just got the update brief from the S2 (the player's guide I have calls the update briefing "Step 1"). As we feared, Regional Highway 2, the route toward the nearby city from the airfield we've secured, is congested by refugees. That's okay, I'm on it. It is my turn to present, and now we're in Step 2 and I'm refining my briefing for the S3. The action I'll be briefing is a multi-capability effort designed to clear the route so the MEU can advance unimpeded while simultaneously directing local Montanyan civilians to a disaster relief site the civil affairs folks are establishing on the other side of the airfield (by the civilian terminals, and accessed by a different route than the one the MEU intends to use to advance).

Time is called to start Step 3, and we all stand as the S3 role-player comes into the room. Since I'm briefing first, I start with a quick review of the situation and orient everyone to the BLUE team game board. I point out the airfield, where MEU elements are at or near the airfield, and the big blue arrow showing the intended line of advance. Then I point out the cluster of civilian icons along that route and note that this is the problem this action is intended to address. I also point out where the relevant information-related capabilities are: the icon for the civil affairs detachment and their relief center, the icons for the military information support operations (MISO) loudspeaker teams and the mobile broadcasting capability, and back on the ship the location of the production shop that will produce the leaflets I'm going to propose we drop.

With the stage set, I then describe how we're targeting two audiences: civilians who have evacuated and civilians who have remained at home along and adjacent to Regional Highway 2. Radio messages, loudspeaker broadcasts, handbills, and leaflets, all in the Montanyan language, will instruct civilians to shelter in place to avoid injury from fighting in the area, and will instruct those who cannot remain in place or those who need medical assistance to proceed to a displaced persons and aid site on the north side of the airport, with instructions regarding the roads to use to get there. I have mock-ups of the leaflets and handbills, and a storyboard for the radio and loudspeaker broadcasts.

I finish by explaining that the desired end state is dramatically reduced traffic on Regional Highway 2, with a secondary effect of increased civilian awareness of and activity at the civil affairs relief site. I list measures of performance (MOPs), which include number of leaflets dropped in the target zone, hours of radio broadcast, and hours of loudspeaker broadcast. Measures of effectiveness (MOEs) should be observable by reconnaissance assets or through overhead imagery, and also reports from executing units: What do the loudspeaker detachments see happening when people hear their messages? What level of activity does the civil affairs detachment observe?

The S3 nods approvingly, and then asks if there is time to get all of these MISO products approved.

It is a good question. I say that yes sir it is all good to go, as all the messaging is within the bounds of pre-approved MISO products and series. MISO always has a standard set of pre-approved series aimed at civilians: "stay" messages (shelter in place), "go" messages (evacuate, with instructions on where to evacuate), and messages about humanitarian relief and where to go to get it.

The S3 then asks me if I have my three reasons this action will succeed ready to go for the engagement step. I do have them ready, and I share them.

The S3 approves my action. After a brief from another team member, the S3 approves a second action as well, then departs. We have about 15 minutes left before we'll head to the engagement room for Step 4. We take turns trying to think of stuff the other team might say in their rebuttals, stuff about why our actions might not work, and things we might say in our counterarguments.

As we enter the engagement room, there is an excited hush. The engagement game board is a giant laminated plotter map that covers most of the floor. The judges sit behind a table at the end of the room, flanked by the S2 and S3 role-players for both sides. The timekeeper (who operates the big countdown clock), the narrator, and some other EXCON personnel (including the GREEN representative) sit at another table perpendicular to and on the right of the judges. Our whole team sits in chairs along one wall, and the other team sits along the other open wall.

The head judge calls engagement to order. The S2s both stand and give a quick overview of the current location of both sides' forces and the current status of the operation. It is always interesting to see the hybrid view of both RED and BLUE situational awareness. I know that "what happens in the engagement room stays in the engagement room," but it is still interesting to see where the RED forces actually are and what their maneuver intent for the turn is.

The head judge calls the team leaders up to determine the order of actions. The head judge has had input from both S3 role-players so knows at least the broad outlines of the actions we'll present. Apparently, none of the actions are head-to-head and there are no special sequencing requirements, so we'll present our actions in alternating sequence depending on who rolls highest. Our team lead rolls highest, so my action, action BLUE 1, is first.

The timekeeper starts the clock, and it begins to count down from 5 minutes. As the clock ticks down, I lay out the details of the action pretty much the same as I did for the S3, and I pass the mock-ups up to the judges. I stand on the plotter map and use a pointer to show where the different capabilities will be, and where the different drop zones and broadcast areas are. I finish well before my time expires with the three reasons this action will succeed. First, I note that the people are scared and confused and that these products offer trustworthy information and clear guidance and so should be effective. Second, I note that the two behaviors we're promoting (stay in place or head to a relief site) are probably things many members of the target audience were considering doing anyway, and our messaging just provides more information and instructions regarding that decision. Third, I note that the action will succeed because so many members of the target audience will get the message, sometimes repeatedly and through multiple media: We've got a bunch of different communication modes that we're employing.

I thank the judges and sit down. My team lead flashes me a "thumbs up." The timekeeper resets the clock to 4 minutes, and the head judge instructs the RED team to prepare their rebuttal. They circle up and begin to discuss furiously. We huddle up as well and discuss some of the possible lines of rebuttal the other team might use.

The timekeeper calls "time" and resets the clock to 2 minutes, beginning the rebuttal countdown. The RED team representative stands and offers three reasons they think the action will fail. First, she notes, the messages are all clearly from the U.S. force, and the United States is not a credible messenger in this area. Second, even if this action makes some people shelter in place and draws some away to the north side of the airfield, the general level of panic among the civilian population is still going to leave Highway 2 too congested to allow the MEU to advance. Third, she says, she and other teammates know that MISO messages take a long time to produce and approve and asks how long we've been planning on this action, and if that includes enough time for product approval?

I stand to launch my counterargument, but before I start the timekeeper waves me back to the team and reminds me that we have 2 minutes to discuss our counterarguments before I make them. The timer starts ticking down from 2:00. It doesn't take us long. We discuss for less than a minute, and I then announce, "I'm ready." The timekeeper resets the clock to 1 minute, starts the timer, and says, "Proceed with counterarguments." I offer my two counterarguments, noting that in fact our intelligence summary indicates that the United States is well liked and well respected in this part of Montanya, so messages from U.S. forces should be viewed as highly credible. And, even though this action is a relatively new development in our plans, based on observing displaced persons clogging roads, as I told the S3, MISO always has messages of this sort as part of their pre-approved materials. I retake my seat.

The judges briefly confer, making notes on their scoresheets. The head judge asks the room, "Well, which is it? Is something like this part of a pre-approved series, or not?" Several of my team-mates nod and say, "Absolutely," but it is firmly sealed when one of the other EXCON personnel, a MISO subject-matter expert, confirms that such products are typically pre-approved. One of the other judges asks about timelines for developing, printing, loading, and distributing leaflets. I don't know the answer, but one of my teammates does, and his timeline is confirmed by that same MISO expert. The head judge turns to the GREEN representative on EXCON. The GREEN rep says he doesn't have any questions but that he wants to confirm that the United States is well regarded in this region of Montanya, so messages from the USMC should not be viewed as inherently lacking credibility.

The judges confer and work on their scoresheets and we talk quietly to each other; the debate seems like it went our way, and so did the judges' questions and observations. Maybe we'll get a nice, low target number.

The judges complete their scoresheet and the head judge signals for attention. The head judge announces that, after considering the difficulty of achieving the effect desired, the quality of planning demonstrated, and the outcome of the debate, the consensus target number for this actions outcome roll will be seven. The whole team is pleased. Seven is the lowest target number we've had for any action so far, so we have a really good chance of success.

I step forward toward the dice. My action, my chance to roll. I roll all three dice. They come up 4, 4, 2—a total of 10. Ten is not that great a roll, but it is good enough here, as it beats the target number by 3. The judges check their sheets to compare the roll total with the margin of success they identified for this action. The head judge announces, "Full success." The narrator begins an explanation of events in which the delivery of all the different messaging products goes off without a hitch, and within a few hours, the number of civilians on Highway 2 decreases, with the general flow of traffic moving more toward the north side of the airfield. Although there is still some congestion on Highway 2, he says, the MEU will be able to advance as planned.

Step 4 proceeds with a RED presentation, then our team's other action, and finally RED's final action. Our other action and one of RED's actions both fail, but RED gets a partial success with a fake atrocity video and is starting to build growing indignation both back in Centralia and in the international community. I'm already starting to think of things we might do to help head that off.

In Step 5, the narrator summarizes all of the con-sequences of the actions and wraps up the maneuver progress for the turn. The MEU is able to advance into the outskirts of the city, meeting the commander's objective for the turn, but it is clear that the atrocity video is going to demand our attention next turn.

2. DETAILED PROCEDURES FOR THE CONDUCT OF THE FIVE STEPS OF THE IWX WARGAME

This section provides detailed instructions, procedures, and rules for the five steps of this wargame.

2.1. Detailed Conduct of Step 1: Receive Scenario and Situation Update

2.1.1. Overview of Step 1

During Step 1, teams receive information on how many actions will be allowed in the turn, an update on the current state of the operation; how (if at all) it is deviating from the planned scheme of maneuver for their side; the current locations of available IRCs; the status of GREEN; and any updates on injects or changing scenario conditions. The Step 1 overview summarizes the current state and lays out the intended maneuver progress for the turn for each side. Step 1 takes place simultaneously in each team's planning space.

2.1.2. Inputs to Step 1

During Step 1, players will be updated on changes to the operation and environment; on expected events in the next turn; on the relevant status of GREEN; on the current location and status of available IRCs; and on timing information (both in the scenario and deadline/events in the game schedule) for the coming turn. In the first turn of the game, players should know the background of the context and how the operation is slated to commence. Current status of the operation is "beginning" in turn 1, and in subsequent turns should have been covered in Step 5 of the previous turn, but EXCON personnel providing updates as part of Step 1 should also be able to answer questions about current state or outcome of the previous turn as needed. The following will be needed for Step 1, all of which must be provided by EXCON:

1. game boards (either physical, digital, or both) and appropriate tokens, markers, or other symbology

2. game board positions depicting the current state, as known by each side

3. game board positions depicting either the intended maneuver and other efforts planned for this turn *or* the intended end state for this turn, for each side

4. game board positions for each of the IRCs available to the side

5. game board depictions of major GREEN elements and expected GREEN activity

6. a briefing or notes for a verbal update for each side detailing:
   - the maximum number of actions that each team will be allowed to take this turn
   - what amount of time in the scenario the coming turn will represent
   - intended maneuver and other efforts planned by the friendly side for the turn
   - intended end state for the friendly side for the turn
   - any deviations to the original friendly maneuver plan for the turn
   - most likely/most dangerous course of action information regarding the other side
   - available IRCs and where they are located
   - relevant expected GREEN actions or status
   - any injects that will occur during the turn (these should take the form of inject cards; see Sections 2.6.1.2 and 3.4.6)
   - deadlines and expected event times (where the players need to be and when) for the remaining steps in the turn.

Ideally, all of these materials will be preplanned/prepackaged for several possible storylines before the game and will only need to be adjusted based on variation in operational outcomes as necessitated by extreme success or failure by one of the teams of players or due to optional injects.

This update is provided by EXCON. If assigned as a role, the S2 role-player should take the lead for this presentation.

OPTIONAL: The S3 role-player may attend the update and may present a portion of the update.

OPTIONAL: The GREEN portion of the update brief can be presented by a separate EXCON member responsible for GREEN (under this option, the GREEN representative would need to travel to each team room separately).

Game boards should be managed by a game board keeper (if such as been designated) or jointly by the S2 role-players. EXCON lead has responsibility for determining and announcing the number of actions each team will be allowed in the current turn.

OPTIONAL: EXCON lead determines the number of actions, but the number is announced by the S2 role-player as part of the update briefing.

OPTIONAL: The maximum number of actions per turn is not announced to the players but is instead told privately to the S3 role-players, and they run their approval process in such a way as to only allow the desired number of actions to actually be approved.

OPTIONAL: If time permits, the hard cap on maximum number of actions may be removed (see Section 2.1.2.2).

2.1.2.1. Considerations for Setting the Number of Actions per Turn

The number of actions allowed per turn may vary in different IWX wargames and *it may vary in different turns of a single game.* The maximum number of actions allowed in a turn is usually either two or three per team.

Playtesting revealed that more actions consume more time both in preparation (Step 2) and especially in adjudication (Step 4). Where action counts were higher (three or even four per team), quality of proposed actions tended to decline and adjudication time in Step 4 ballooned. However, greater numbers of actions create more opportunities for something to happen in the game and for players to experience the joy of a successful action. Too few total actions accompanied by poor dice luck by a team could result

in multiple turns passing without a successful action for the team.

Having a declared maximum number of actions per turn is an artificiality of the wargame. During an actual operation, the number of actions (or equivalent) that an OIE OPT might plan or direct is constrained by the availability of IRCs (and their throughput), the size and bandwidth of the OPT itself, and the time available for planning and preparation prior to the beginning of the operation. For a wargame, another consideration is that time constraints limit the number of actions. In deciding how many actions to allow, whether to allow any flexibility on that number (hard cap or soft cap), and whether to publicly announce the allowed number of actions or use the S3 in Step 3 as a forcing function to control the numbers of actions approved (see Section 2.3.4), EXCON needs to recognize that tradeoffs are being made between realism (realistically actions are constrained by available resources, not a fixed number) and game practicality (practically, Step 4 is the most time-consuming step in the game and increases substantially in length for each additional action adjudicated).

OPTIONAL: Continuing actions may be considered as not counting against the action limit.

2.1.2.2. OPTIONAL: Removing the Hard Cap on the Number of Actions per Turn

If time permits, the hard cap on maximum number of actions may be removed, such that teams may be able to take four or even five actions in a turn. An S3-driven "soft cap" on actions preserves verisimilitude better than an arbitrary hard cap of two or three actions per turn as a strict game rule. However, even if extra time is available, if more than two or three actions are allowed, they all should be very tight. If an S3-driven "soft cap" is being used, the S3 should raise the bar for approval for additional actions. See Section 2.3.4 for suggestions on how to do this.

2.1.3. Player Activities During Step 1

In this step, players receive updates and information and take notes on new information as needed. Players ask clarifying questions about the updates they receive.

2.1.4. EXCON Activities During Step 1

EXCON provides the updated game board and the briefing or verbal presentation of updates, changes, number of actions allowed for the turn, and injects (injects are described in Sections 2.6.1.2 and 3.4.6). EXCON provides these updates to both teams separately but simultaneously, so the absolute minimum required EXCON personnel for the step is two. If available, these presentations should be made by a role-played S2. Ideally, the S2s should each be accompanied by a notetaker.

Ideally, two or (better still) three EXCON personnel should be present for the update to each team in order to have more expertise available to answer any questions that emerge and to take notes about questions (and the answers given). The total recommended EXCON personnel requirement for the step is between four and six.

> OPTIONAL: The S3 role-player may attend the update and may present a portion of the update.

> OPTIONAL: The GREEN portion of the update brief can be presented by a separate EXCON member responsible for GREEN, with that individual visiting both team rooms.

EXCON lead has responsibility for determining and announcing the number of actions each team will be allowed in the turn to come.

> OPTIONAL: EXCON lead determines the number of actions, but the number is announced by the S2 role-player as part of the update briefing.

> OPTIONAL: The maximum number of actions per turn is not announced to the players but is instead told privately to the S3 role-players, and they run their approval process in such a way as to only allow the desired number of actions to actually be approved.

> OPTIONAL: If time permits, the hard cap on maximum number of actions may be removed (see Section 2.1.2.2).

> NOTE: EXCON should not necessarily answer any and all questions asked by the players. Questions about game procedures should be answered fully. But requests for information (RFIs) about past or planned future events should be answered only to the extent that an OIE OPT would have access to such information as part of situational awareness and other intelligence flows within that headquarters within the scenario. EXCON should also be careful with the invention of details that players should have access to but are not in the materials provided, as these inventions must then be incorporated into the ongoing wargame scenario if they prove to be important details. Any RFIs answered by any EXCON representative during Step 1 or at any other time should be recorded on an RFI matrix (notes indicating who asked what, when, and what they were told) and submitted to EXCON central so that any changes in the scenario world implied by the answer can be recorded and carried forward throughout the game.

2.1.5. Outputs from Step 1

Step 1 should produce (1) the notes the players take to support their efforts in later steps and (2) the notes EXCON takes based on any exchanges of information with the teams/players.

2.1.6. Time Allowed for Step 1

Step 1 should be completed in 15 minutes. Step 1 may be unnecessary for the first turn of the game. Time allocated to Step 1 might serve as overflow for Step 5 on the previous turn, should events be behind schedule or should results and reset take longer than anticipated.

2.2. Detailed Conduct of Step 2: Prepare to Present

2.2.1. Overview of Step 2

In Step 2, teams revisit their plans as needed and prepare their actions for the turn. This step takes place in each team's planning space. Teams compose an oral presentation for the S3 role-player for their side (for Step 3) and also marshal their arguments supporting these actions for engagement (Step 4). Teams prepare mock-ups of any products that will be employed in the actions. The presenting player for each action rehearses their presentation and makes sure they are ready to offer compelling explanations and arguments.

2.2.2. Inputs to Step 2

At the outset of Step 2, players should have all of the inputs they need. These should include their existing plans; their understanding of the capabilities available to them or that they can request; all the scenario background materials; and their understanding of the current situation and the desired progress of the operation during the turn, the status of relevant GREEN groups, and the locations and status of available IRCs, all based on their notes. Other possible inputs might include answers to any questions that the players come up with as they prepare, limited as above to answers to which they should be entitled (game procedural answers, or answers that an OPT in the HQ they represent would be able to get within the confines of the time represented by the turn).

2.2.3. Player Activities During Step 2

During Step 2, players update their overall plan as needed in response to changes in the anticipated flow of the operation, unanticipated actions by the opposed side or team, and other changes in the scenario context. Players also prepare their actions for the turn. If an EXCON member is present, teams may ask questions or refer RFIs through that individual. If no EXCON representative is in the team room or

if they do not know the answer to a posed question, teams may send their team leader to the EXCON control room to ask questions as needed.

2.2.3.1. Preparing Actions to Present

2.2.3.1.1. How Many Actions to Prepare

Teams are told during Step 1 how many actions they may present in Step 4 (see Section 2.1.2). Teams may wish to prepare more than the maximum allowed number of actions to present for approval in case one or more actions are not approved. Players might also plan ahead and begin to prepare actions for later turns. Only actions approved in Step 3 may be taken in Step 4.

Actions begun in a previous turn and continued in a later turn must be presented again for approval in Step 3 and presented for adjudication during Step 4 in that later turn. Action previously attempted that failed may also be re-presented for approval and re-attempted.

> OPTIONAL: Continuing actions may be considered as not counting against the action limit.

> OPTIONAL: The maximum number of actions per turn is not announced to the players but is instead told privately to the S3s, and they run their approval process in such a way as to only allow the desired number of actions to actually be approved.

> OPTIONAL: If time permits, the hard cap on maximum number of actions may be removed (see Section 2.1.2.2).

2.2.3.1.2. Rotation of the Presenting Player

Each team action is presented by a single player. That player is the presenting player for that action and must be the one who presents the proposed action in Step 3. If the action is approved, that player must be the one who presents the approved action for matrix discussion and adjudication in Step 4. If the action continues into later turns as a continuing action, that same player must continue to present the action.

Players must rotate presentation of actions through the entire team to ensure that each player has the opportunity to present, and so that exercise staff have an opportunity to observe each player's demonstrated learning. A player who has already served as the presenting player for an action may not be the presenting player for another action until all other players on the team have served as the presenting player. This rotation continues across turns and phases.

2.2.3.1.3. Materials to Present a Candidate Action

The presenting player must be prepared to verbally explain and justify a proposed action to the S3 role-player in Step 3. To present a candidate action, players should have prepared oral remarks. If they wish, players may enhance their oral presentation with slides or support it with tokens or other representations to be placed on the game board (or moved on a digital game board). Action proposals must detail the following:

- target, TA, audience, and location
- effect desired, including the location, time, and duration of the effect
- purpose of the action and how that purpose contributes to the commander's intent and the scheme of maneuver for the phase
- forces or capabilities that will conduct or contribute to the action, where they will be located (on the game board), how they will get to that location, and when their activities will occur
- the desired end state, including what failure will look like and what success will look like
- an assessment plan, including measures of performance and measures of effectiveness, how those measures will be collected/observed, and how long after the activities/effects measures will be reported
- a list of at least three reasons the action will succeed (only three reasons can be presented in Step 4, but more might be presented in Step 3).

In addition, teams will need to prepare (and the presenting player will need to present) mock-ups of **any products that the action requires**, such as leaflets, radio scripts, storyboards for video production, key leader engagement (KLE) talking points, or other

artifacts that would support assessment and adjudication of the action should it be approved.

The presenting player should rehearse their presentation during Step 2, time permitting, so that other team members can provide them with feedback and to maximize their preparedness for Step 3.

2.2.3.1.4. Order of Presentation

Players should plan to present their actions in the order in which those actions are scheduled to begin, with continuing or ongoing actions presented first (see Section 2.2.3.1.5). Players should present actions in the same order in both Step 3 and Step 4 unless instructed to do otherwise. EXCON may require a change in order of presentation during Step 4 in order to prevent contradictions or to align directly opposed actions for head to head adjudication. See more on order of presentation in Section 2.4.3.2.

2.2.3.1.5. Repeated, Recurring, or Ongoing Actions

Players may wish to conduct actions from previous turns again in later turns, either because the action is continuing or because the action previously fell short and they want to try it again. If a team wishes to repeat or continue an action, they need to present it again for approval in Step 3 and present it as part of engagement in Step 4. If little has changed, then repeat presentations may be somewhat abbreviated. However, if something significant about the environment or execution is different, the team should make a full presentation of the action.

As noted, a player who is designated as the presenting player for an action remains the presenting player for that action, no matter how many times the same action is presented.

2.2.3.1.6. Re-Presenting a Previously Disapproved Action

For actions that the S3 role-player disapproved in a previous turn but recommended for further development, the presenting player remains the same; that is, if an action is not approved during Step 3 in one turn and is presented for approval again in a subsequent turn, it must be presented by the same presenting player.

2.2.4. EXCON Activities During Step 2

EXCON engagement with the players is minimal in this step, as the step is primarily a planning, preparation, and rehearsal period for the teams. EXCON might have four roles in this step: (1) being available to answer questions that come up, (2) providing mentorship, instruction, or advice to players, (3) observing the teams to see what they are going to be proposing as actions, and (4) using those observations of planned actions to prep the S3 role-player, to begin thinking about how such actions might be scored on scoresheets during Step 4, and to begin planning for how such actions might impact the flow of the operation and the storyline (for Step 5 of the current turn and Step 1 of the next turn).

Step 2 requires a minimum of one EXCON person available to answer questions located in the EXCON control room. It is preferable to have at least two additional persons available to take notes and to observe the teams. EXCON personnel required for the step range from one to three, with three being ideal.

OPTIONAL: If available, the S2 role-players may remain in the team rooms during Step 2 to answer RFIs and offer advice. However, the S3 role-player **should not** remain in the team room and advise, as doing so could compromise their credibility in the role of hard-nosed S3 and their approval or disapproval of actions in Step 3.

OPTIONAL: Instead of seeking to maintain the hard-nosed credibility of the S3 as a challenge/barrier in the process, the game could instead employ the S3 role-player as a mentor, offering advice to the team and trying to shepherd their actions to success. In this case, extra care must be taken in Step 3 to ensure that substandard actions are not approved.

NOTE: Any RFIs answered by any EXCON representative during Step 2 or at any other time should be recorded on an RFI matrix and submitted to EXCON central so that any changes in the scenario world implied by the answer can be recorded and carried forward throughout the game.

OPTIONAL: If the optional rule omitting Step 3 is in use (see Section 2.3.7), then EXCON will need to make personnel available to adjudicate the availability of permissions and/or capabilities from higher echelons (as described in 2.3.4.2) during Step 2.

2.2.5. Outputs from Step 2

Step 2 should produce a set of candidate actions ready to be presented to the S3 role-player in Step 3. These candidate actions should be composed of some or all of these elements: notes for an oral presentation; a supporting briefing or plan to use markers, tokens, or symbols on the game board; and mock-ups of any products to be used as part of the action.

EXCON should gather these materials as outputs for use in preparing for steps 4 and 5. This might involve EXCON members (the S3s and/or the judges) viewing materials in a team's folders on a shared drive, or an EXCON notetaker photocopying some of a team's materials or requesting that the team print an extra copy of working materials and then share them with the relevant S3 and/or the judges. EXCON outputs include a prepared S3 role-player who is aware of and ready for the actions that will be presented based on information passed from the EXCON notetaker or other EXCON observers or viewed on shared drives.

2.2.5.1. Battle Drills

An additional possible output from Step 2 is one or more possible battle drills: things the side will do if certain conditions (usually bad conditions) come to pass. Such battle drills probably should be part of the larger plan prepared during initial planning portion of IWX. Although such battle drills are likely not formal "actions" as understood in this wargame, they might be useful as arguments in the rebuttal or counterargument stage of Step 4. "Counteractions" and other unplanned responses to the actions of the other team are explicitly forbidden from being used in rebuttal or counterargument *unless* such actions are part of a battle drill that is documented in the team's plan and can be verified by the role-played S3. Having such battle drills (and sharing them with the S3) is not required, but they certainly could prove useful in the game (and in actual operations).

2.2.6. Time Allowed for Step 2

Step 2 should require 30–60 minutes to complete. If events are unfolding in a way that is consistent with a team's plans, then it should not take the team long at all to identify elements of their plan to propose as actions and prepare presentations for. However, if results have been going against a team or opposed play has caused events to turn in unanticipated directions, players may be scrambling to adjust their plan (and the corresponding game actions) to compensate. Recommended allocation of time for this step is 60 minutes either to allow players some slack time or to allow them sufficient time to adjust and recalibrate. If later turns in later phases of the operation represent reduced amounts of time in the scenario, reducing the allowed preparation time to 45 or 30 minutes might make sense.

> **NOTE:** Players can and should work on elements of Step 2 (and future Step 2s) during slack time, including excess time in other steps, time between steps, time between turns, and time before or after the formal exercise time for each day is complete. For example, players who are not the presenting players for a turn and who are not needed to help presenting players prepare could use Step 2 from one turn to prepare actions (or product mock-ups) for presentation in later turns (accepting that events may force later changes in some of the details).

2.3. Detailed Conduct of Step 3: Present Actions for Approval

2.3.1. Overview of Step 3

Step 3 takes place in each team's planning space. For each proposed action, the presenting player (and only the presenting player) briefs the S3 role-player on the audience and intended effects, capabilities involved, and details of (planned) execution. A different presenting player presents each candidate action. The S3 role-player approves or disapproves each action. Approved actions that require capabilities or permissions not organic to the team's side's forces will be immediately adjudicated as able to proceed or unable to proceed by EXCON based on a special outcome

roll. On the S3 role-player's return to the control room, the S3 role-player back-briefs the judges on the approved actions, and the judges review approved actions and begin to complete action scoresheets in preparation for Step 4.

2.3.2. Inputs to Step 3

Step 3 has numerous requirements from both players and EXCON. Teams must provide two or more candidate actions ready for presentation and including all of the elements listed in Section 2.2.3.1.3, each to be presented by a different presenting player. EXCON must provide a prepared S3 role-player who knows the current state of the operation, the progress intended for this turn, and the number of actions allowed for the turn (which may or may not be known to the players; see Section 2.1.2) and who is ready to be critical of the presented plans but is also prepared to make realistic approval or disapproval determinations.

The S3 role-player should have a number of blank proposed action approval checklist forms equal to the number of actions to be presented. Ideally, the S3 role-player will have some sense of the candidate actions they will be asked to approve, either from familiarity with the team's plan or from cueing from EXCON observers dedicated to that team. As discussed in Section 2.3.4.2, EXCON should be prepared to adjudicate any actions that cannot be approved by the S3 alone.

2.3.3. Player Activities During Step 3

Players present candidate actions to the S3 role-player for approval. Players present the actions in the order in which the actions would begin. Continuing or ongoing actions are presented first. Presentations should include all of the elements listed in Section 2.2.3.1.3 and repeated below. Each action will be presented by the presenting player. Unlike the presentations in Step 4 (when *only* the presenting player may address the judges), in Step 3 the presenting player must take the primary role in the presentation, but other players on the team are allowed to help in answering any questions the S3 role-player may have.

FIGURE 3

A Player Proposes an Action to the Role-Player S3 During IWX 20.2

Photo credit: Nate Rosenblatt, RAND.

If an action is a repeat action from a previous turn (either because the action is ongoing, the action was not previously approved, or the action was previously attempted but was unsuccessful), then the same presenting player who previously presented the action must present it.

Players will take notes on the feedback they receive from the S3 role-player that will help them refine their presentation of the approved actions in Step 4, or help them refine actions that were not approved should they wish to propose improved versions of those actions in later turns.

After receiving approval for their actions for the turn, players will make any final adjustments to the presentation of the actions (to include narrowing down the reasons that the action will succeed to exactly three reasons) for Step 4.

2.3.3.1. Materials to Present a Candidate Action (compressed from Section 2.2.3.1.3)

To present a candidate action, the presenting player should either have slides or notes for a verbal presentation supported by tokens or other representations to be placed on the game board that will allow them to detail the following:

- target, TA, or audience, and location

- effect desired, including the location, time, and duration of the effect
- purpose of the action and how that purpose contributes to the commander's intent and the scheme of maneuver for the phase
- forces or capabilities that will conduct or contribute to the action, where they will be located (on the game board), how they will get to that location, and when their activities will occur
- the desired end state, including what failure will look like and what success will look like
- an assessment plan, including measures of performance and measures of effectiveness, how those measures will be collected/observed, and how long after the activities/effects measures will be reported
- a list of at least three reasons the action will succeed (only three reasons can be presented in Step 4, but more might be presented in Step 3).

In addition, the presenting player will need to present mock-ups of **any products that the action requires**, such as leaflets, radio scripts, storyboards for video production, KLE talking points, or other artifacts that would support assessment and adjudication of the action should it be approved.

2.3.4. EXCON Activities During Step 3

Each S3 role-player will meet with their respective side's team and receive a briefing on their proposed actions. Each action will either be approved or disapproved. Approved actions can be unconditionally approved for presentation in Step 4, or conditionally approved pending certain revisions before Step 4. Disapproved actions will either be permanently rejected or conditionally rejected with advice to revise the proposal for a following turn.

Each action must be presented by a different presenting player. A player may not repeat as the presenting player until all players have presented. The exception is for an action that was previously presented (either because it was disapproved, was approved and failed, or was approved and is continuing), in which case it must be presented by the same presenting player who previously presented it. If an action was previously approved or is continuing with details unchanged, the S3 may elect to ask the presenting player to review all of the details as a full action proposal, may request a summary review, or may simply declare that the action continues to be approved.

> **NOTE:** One player must be the presenting player and make the whole presentation, but the presenting player may confer with other team members in response to S3 role-player questions, especially when IRC expertise is distributed across the team.

The number of approved actions allowed for each team per turn is determined in Step 1 (see Section 2.1.2 and also 2.2.3.1.1). The number of allowable actions per team for the turn may be announced to the players, or it may be told privately to the S2s. **S3s should approve no more than the maximum number of actions allowed for the turn.** If the team has more actions available than the number allowed, then the S3 may have to choose from the available actions (or force the players to choose). It is up to the S3 role-player to handle this. If the number of allowed actions has been announced, it will be easier; if the turn allows two actions, then the S3 could just ask the team to start with their best two actions, and if one of them for some reason cannot be approved, the S3 could then ask the team to present an additional action. If the maximum number of allowed actions has intentionally not been publicly announced, then the S3 must find some other way to approve only the allowable number of actions (perhaps by cutting the meeting short, or perhaps by acting unreasonably or asking unreasonable questions).

The S3 should approve a number of actions equal to or less than the allowable number of actions for the turn. Another consideration that the S3 could raise to influence the approval process is available resources. How much of each IRC is required to do all of the things the team is proposing? What level were those capabilities at at the start of the turn? How much lift or other transportation is available? How much of an additional force protection burden does the action impose? These kinds of considerations can be used to force teams to prioritize and choose between actions or otherwise keep a limit on what could become too great a number of actions.

The S3 should use *GA6: S3 Role-Player's Action Approval Checklist* (see Section 2.3.4.1) to help their evaluation of actions. Regardless of the number of actions allowed per turn, all actions must still meet a minimum standard of quality. **As S3, do not approve substandard actions!**

> OPTIONAL: The maximum number of actions per turn is not announced to the players but is instead told privately to the S3s, and they run their approval process in such a way as to only allow the desired number of actions to actually be approved.

OPTIONAL: As discussed in Section 2.1.2.2, if time permits, the hard cap on maximum number of actions may be removed. However, even if extra time is available, if more than two or three actions are allowed, they all should be very tight. If an S3-driven "soft cap" is being used, then the S3 role-player is the game's only realistic mechanism for controlling number of actions, and in Step 3 this person should raise the bar for approval for additional actions. This could be done in various ways: The S3 role-player might ask more and harder questions of the presenting player. The S3 role-player might express greater skepticism and perhaps indicate scarcity of assets under the S3's control. For example, the S3 role-player might say, "Well, I can enable security and mobility for action #2, or I can do it for action #4, but not for both, and I prefer action #4, so #4 is approved and #2 will have to wait until later." As needed, the S3 role-player can also deny further actions for a given turn by asking questions for so long that the meeting time is exhausted, or by simply ending the meeting early on the grounds that S3s are busy and in demand, thus tabling discussion of additional actions until later.

The S3 role-player should make sure that the starting position of all IRCs is properly noted on the game board and that presenting players refer to which capabilities will be used and how they will move from place to place as required. If the same asset needs to be in two places at the same time or an asset is required to travel an unrealistic distance in an impractical manner, then the S3 should use that to force the players to adjust their actions.

Step 3 requires a S3 role-player for each team (two total), and should include a notetaker/record keeper for each S3 (two total). Thus, this step requires between two and four EXCON personnel across the two team rooms.

OPTIONAL: Required personnel can be reduced if the S3 is responsible for taking their own notes, but they have a lot of responsibilities to keep track of in this step already.

2.3.4.1. S3 Role-Player's Action Approval Checklist

Approval of actions is a holistic process that can be based on the S3 role-player's "gut." However, to help structure that decision, and to help the S3 role-player provide constructive feedback likely to help players improve their later presentations, we provide a checklist in the downloadable game materials as *GA6: S3 Role-Player's Action Approval Checklist*. **We strongly encourage the S3 role-players to use the checklist.**

NOTE: A proposed action does not necessarily have to fully meet all checklist criteria in order for it to be approved; that is left to the S3's discretion (and "gut"). Review of the approval checklist is an opportunity for the S3 to serve as a forcing function for completion of all presentational elements. While an S3 may approve an action that does not include all elements, they should at least ask after missing elements and encourage the players to add or expand upon those elements. If an element isn't in the Step 3 presentation, it likely won't be in the Step 4 presentation unless the S3 takes steps to ensure that it will.

GA6: S3 Role-Player's Action Approval Checklist includes two elements for each criteria: (1) a check as to whether or not the elements were included in the presentation (for example, did the presenting player describe the TA and indicate where on the game board they are expected to be?) and (2) a go/no-go assessment of that element of the proposed action (for example, yes, there is a specified TA, but is it a reasonable audience to target?). The checklist includes the following items:

- target, TA, or audience is described, including location
- effect is described, including intended location, time, and duration
- purpose and connection to maneuver are described
- capabilities/forces that will contribute are described, along with where they need to be, when, and how they will get there
- any products involved are furnished as mock-ups

- desired end state is described, including a description of success and a description of failure
- assessment plan is described, including MOP, MOE, and sources
- at least three reasons are offered for the success of the proposed action.

2.3.4.2. Actions That Require Capabilities or Permissions That an S3 at This Echelon Could Not Approve

Some proposed actions may require capabilities outside the organic capabilities indicated as available to each team, and some proposed actions may require approvals or permissions at higher levels than the S3 at this echelon. If such approval or release of additional capability is pro forma or all but guaranteed, then the S3 role-player may approve the action, indicating that although they themselves cannot approve or release, it is sufficiently automatic that EXCON immediately makes a favorable adjudication. Similarly, if approval is completely out of the ques-

tion, or if a real S3 wouldn't even give the OPT permission to ask, or if the approval timeline is longer than the time allowed within the turn, then the S3 can immediately reject the action.

If, however, there is genuine contingency and realistic uncertainty about whether certain capabilities will be made available from elsewhere in the force or permission given from a higher echelon, this can be resolved with an outcome roll during the presentation of actions to the S3. Players should still present their action as if it would be approved; then, immediately following S3's approval decision, an outcome roll can be used to determine whether necessary capabilities and permissions are available.

To use an outcome roll to adjudicate availability or permission from higher up, first establish a target number using Table 2 as a rough guideline. Basically, if capabilities or permission is likely but not a sure thing, then the target number is 7; if availability or permission is truly uncertain or contingent, like a coin toss, then the target number is 10. If availability or permission is unlikely but still possible, then the

TABLE 2

Example Starting Target Numbers for Availability of External Capabilities or Permissions

Requires out-of-echelon capabilities:	Sample Target Number
The desired capability is present in theater and is commonly but not always available	7
The desired capability is present in theater but is a somewhat busy capability and could easily be unavailable	10
The desired capability is low density/high demand and may need to come from outside the theater or is otherwise hard to get	14
Requires permission or approval from a higher level:	Sample Target Number
Permission or approval could come from only a few echelons up and the level of risk is low, but there is a chance the approving authority is too busy or will request more information	7
Permission or approval requires a long approval chain, with more voices that might not concur, or risk associated with the action is moderate and rejection based on risk is a real possibility	10
Permission or approval must come from the highest levels but is fairly routine; still possible that approval will take too long or that a high-level stakeholder will be risk averse	14
Risk associated with the action is relatively high, and approving authority may reject the action or seek more thorough risk analysis	14

target number is 14. *GA6: S3 Role-Player's Action Approval Checklist* includes a worksheet for actions requiring additional capabilities or permissions.

The following are examples of possible appropriate modifiers to the outcome roll; other modifiers can be created as desired by the adjudicating EXCON team/S3 role-player:

- OPTEMPO of operation generally slow (not much demand on other assets/HQs): +2
- requested previously but not approved by a thin margin: +2
- requested previously but not approved by a wide margin: –3
- friendly force is under significant duress and supporting or vectoring capabilities are all busy: –1.

> **NOTE:** The way in which outcomes of actions proposed and rejected previously were described can dramatically affect conditional modifiers. For example, if this same action was proposed in a previous turn and failed to meet a permission-based outcome roll by 1 or 2, and the explanation given was "approval came, but it was too late and too slow for you to conduct the action this turn," then on a later turn the roll could receive a strong positive modifier, or the roll could be dispensed with entirely and noted as approved for this turn based on the description in the previous turn.

If a proposed action requires both external capabilities and external permissions, then the S3 role-player can choose to resolve the uncertainty with either a single outcome roll or with two outcome rolls. Be advised that requiring two outcome rolls creates two possible points of failure and thus lowers the overall prospects for success for the action, but two rolls might be appropriate for an action with "easy" availability and "easy" permission. If both external permission and capability are required, perhaps use the higher of the two target numbers. This could be justified: If approval comes from a high level for this action, the capability will surely be made available. This would be an S3 decision and at their discretion.

While the guidelines laid out in this subsection are intended to address permission or capabilities to execute OIE required from outside the purview of the S3, they could also be extended to cover intelligence-collection assets or other requirements for observing and monitoring the outcomes of intended actions. Should the assessment plan for an action require assets or collections from organizations or forces outside the control of the current echelon, a similar process could be used, with an outcome roll determining availability. Alternatively, availability/assignment of intelligence assets could be included in a single roll, with either or both capability availability and higher-level approval.

> **NOTE:** Failure to secure assets to monitor or assess an action may not prevent it from being approved; an action could still be approved and attempted with insufficient assessment at the discretion of the S3. This may result in reduced information about the outcome of the action during Step 4 (see Section 2.4.3.9.2).

2.3.4.3. Final Determination of Approval Status of All Actions

After presentation, discussion, and (if necessary) an outside approval or availability roll, the S3 role-player should provide a clear determination of the approval status of each action to the players. There are four possibilities:

1. The action is approved as is.
2. The action is approved with adjustment before Step 4. (In this case, the S3 needs to clearly state what adjustment is required, and whether the S3 needs to see confirmation of the adjustment before Step 4 begins.)
3. The action is not approved in this turn, but should continue to be refined and should be re-presented in a future turn.
4. The action is not approved in this turn (back to the drawing board).

2.3.4.4. "Backroom" EXCON Activities During Step 3

The main EXCON activity in this step is completed by the S3 role-players assigned to each team. However, the rest of EXCON should not be idle. Using observations made by the EXCON notetaker for each team in Step 2 (if assigned), and **with input from the S3 role-players after they return from** receiving action briefings, the three EXCON adjudication judges should prepare draft scoresheets for each approved action so that they are draft complete and need only be updated based on the actual presentation, rebuttal, and any other elements of matrix discussion in Step 4.

The EXCON keeper(s) of the game boards should also ensure that sufficient tokens to represent possible effects of the various proposed actions are available in the engagement room. Game board keeper(s) should also prepare a hybrid representation of both RED and BLUE situational awareness to display during engagement. This depiction should avoid giving away unknown maneuver positions but should provide enough of RED and BLUE positions (possibly ambiguously) to allow teams to describe the locations of their actions.

2.3.5. Outputs from Step 3

Step 3 should produce a number of approved actions (no more than the number announced/determined in Step 1) for each team. Step 3 may also produce one or more candidate actions that are not yet approved but could be approved in future turns for the team to continue to refine. The approval status of all actions should be clear to the players (as per Section 2.3.4.3).

> **NOTE:** For actions that are not approved but are recommended for further development, the **presenting player** remains the same; that is, if an action is not approved during Step 3 in one turn and is presented for approval again in a subsequent turn, it must be presented by the same presenting player.

2.3.6. Time Allowed for Step 3

Step three has two substeps: (1) the presentation of actions to the role-played S3 for approval and (2) final adjustments and preparations to present those plans in Step 4. Total allocation of time for both activities should be 60 minutes. The players can have whatever time is left after their actions are approved (or denied) by the S3 to finalize the presentation of those actions for the next step. The S3 and the rest of EXCON can use this residual time to prepare for a smooth and efficient Step 4.

2.3.7. OPTIONAL: Omit Step 3

Under certain circumstances, almost all of Step 3 might be omitted. This might be necessary if adequate S3 role-players are unavailable, or it might be attractive to save time by omitting Step 3 and treating the presentation in Step 4 as both the practice of justification of proposed action to staff seniors and as engagement for wargame adjudication.

If Step 3 is omitted, then the process for ascertaining whether approval or capabilities are received from higher echelons (as described in Section 2.3.4.2) should take place during Step 2, with team leaders requesting an EXCON adjudicator to support such determinations as needed.

If Step 3 is omitted, information from Step 3 that informs Step 4—such as information about the pending actions by each team necessary for EXCON judges to prepare draft scoresheets (Sections 2.4.2 and 2.4.3.8), for the EXCON narrator to prepare draft narration of outcomes (Sections 2.4.3.8.3.2 and 2.4.3.9.1), and for the EXCON lead to identify head-to-head actions or other details important to the sequencing of actions (Section 2.4.3.2)—will need to be provided by some other means. This could be accomplished by requiring team leaders to submit single-page summaries of their planned actions by the end of Step 2, or by having someone from EXCON (the S2 role-player, a notetaker, or someone acting as a mentor in the team rooms) report this information back to EXCON central (the judges and the narrator at minimum) by the end of Step 2.

2.4. Detailed Conduct of Step 4: Engagement and Matrix Debate

2.4.1. Overview of Step 4

In Step 4, all players from both teams come together in the engagement space, and teams alternate presenting actions approved by their S3 role-player. The same presenting player who presented actions in Step 3 presents them in Step 4. During engagement, each presenting player also presents three reasons why they believe their action will be successful. The opposing team then has a few minutes to discuss and presents three reasons why they believe the action will be unsuccessful or less effective than the presenting player has indicated. The presenting team then has two minutes to prepare and an additional minute to present up to three counterarguments to the rebuttal. (This is a form of what is called *matrix adjudication* in the wargaming community.)

EXCON will have an opportunity to ask clarifying questions, and then EXCON will quickly confer and finalize scoresheets, giving the presenting player a target number and any roll modifiers accumulated. The presenting player will then roll dice for an outcome roll, which will produce a degree of success or failure for the action. The narrator will describe what happens as a result of the action. Teams alternate presenting actions, engaging in matrix discussion, receiving EXCON input and target numbers, and generating outcome rolls until all actions have been adjudicated.

2.4.2. Inputs to Step 4

In Step 4, both teams and all of EXCON meet in the engagement space. Step 4 is where all the outputs of the previous steps come together. Teams will need presenting players prepared to present and defend their team's actions, possibly including briefing slides or a plan to display the action on the game board. EXCON will need a panel of judges prepared to score and adjudicate the actions (three judges, one of whom should be the EXCON lead). The EXCON judges should have sufficient copies of *GA8: Engagement Scoresheet* and *GA9: Head-to-Head Engagement Scoresheet* available to score every action, as well as *GA10: Target Number Calculation Record Sheet* and *GA11: Action Results Record Sheet* to calculate and record target numbers and outcome rolls for previous actions. A member of EXCON should act as timekeeper.

Ideally, EXCON will already know what actions each team will present and will be prepared to support determination of order of presentation (see Section 2.4.3.2), will already have draft scoresheets preliminarily complete (see Section 2.4.3.8), and will have preliminary notes for narration of different possible outcomes for each action (see Sections 2.4.3.8.3.2 and 2.4.3.9.1). This input should have come from a combination of EXCON observers

during Step 2 and the back-brief from the S3 role-player at the end of Step 3.

> **NOTE:** This step requires full participation by all players and EXCON. It is the most exciting step of the wargame, and, as such, everyone should be available to witness it.

With regard to manning requirements for EXCON, minimum recommended roles assigned are as follows:

- the three judges (one of whom should be EXCON lead)
- the S2 and S3 role-players for each team
- the narrator (who will describe the outcome of adjudicated actions and build results into the scenario storyline; the narrator could optionally be one of the judges)
- a notetaker assigned to each team
- someone with responsibility for IT and network support (ensuring that all presentations have access to and can share any digital materials that are part of their presentation)
- someone to manage the game board (if role-player S2s are used, the two S2s could jointly manage the game board, or this may be a responsibility of the narrator or a narration team)
- someone to serve as timekeeper, managing the stopwatch or timer to constrain times for presentation and rebuttal.

Thus, there are a total of 13 EXCON roles, some of which could be filled by the same individual; for example, a notetaker or S2 role-player could also serve as timekeeper, or the narrator could also be a judge and/or the EXCON lead.

> OPTIONAL: The EXCON judge panel can also include an EXCON representative explicitly focused on and responsible for the GREEN perspective. This could be one of the three judges or could be a fourth judge. If a fourth judge is used, consider also adding a fifth judge to break scoresheet ties (the scoresheet process involves taking medians across all judges' scoresheets, and a median is much easier to calculate across an odd number of judges).

2.4.3. Player and EXCON Activities During Step 4

Because this step involves an iterative sequence of activities that involve both teams and EXCON, these activities are described in a single sequence rather than separated into player activities and EXCON activities, as is the case in other sections of this ruleset.

Table 3 lists the activities in sequence and notes the ruleset subsection in which they are described.

The various presentations, rebuttals, and counterarguments of the teams are timed in order to ensure fairness and to keep things moving. As described in the subsections that follow, the presentation and argumentation of an action uses this sequence and these time constraints:

- Presenting team presents (5 minutes—described in Section 2.4.3.3)
- Non-presenting team prepares rebuttal (4 minutes—described in Section 2.4.3.4)
- Non-presenting team rebuts (2 minutes—also described in Section 2.4.3.4)
- Presenting team prepares counterargument (2 minutes to prepare—described in Section 2.4.3.5)
- Presenting team counterargues (1 minute—also described in Section 2.4.3.5).

> OPTIONAL: Timekeeping times and tracking may be adjusted based on the equipment available and in a matter that suits the preferences of EXCON lead. This may include a "soft limit" where players are told when time expires but are allowed to finish their point.

2.4.3.1. Reorientation to the Game Board and Situation

Before beginning the cycle of action presentation and adjudication, EXCON should set the game board to display a hybrid of RED and BLUE situational awareness and set the stage by reorienting players to the conditions on the battlefield and in the environment at the beginning of the turn, as well as describing major maneuver objectives for each side. This is not intended to be as elaborate as the situation update presented in Step 1 but rather a simple review to remind all participants of the starting positions for the turn. If available, one or both S2 role-players could give this reorientation.

> OPTIONAL: The narrator or the EXCOM lead could provide this reorientation.

TABLE 3
Step 4 Activities

Activity	Time Limit (minutes)	Section
1. Reorientation to the game board and situation		2.4.3.1
2. Determine the order of presentation		2.4.3.2
3. Presenting team presents	5	2.4.3.3
4. Non-presenting team rebuts	4 to prepare, 2 to speak	2.4.3.4
5. Presenting team offers counterarguments	2 to prepare, 1 to speak	2.4.3.5
6. EXCON interrogates, completes scoresheets, prepares for outcome roll		2.4.3.8
7. Outcome roll, outcome determination		2.4.3.9
8. Repeat activities as necessary		2.4.3.1
9. Record all results and prepare for Step 5		2.4.3.12

FIGURE 4

The Map Manager Sets the Game Board Prior to Engagement During IWX 20.2

Photo credit: Nate Rosenblatt, RAND.

2.4.3.2. Determine the Order of Presentation

Teams alternate presenting actions. If either team has a greater number of approved actions to present, that team will go first. If teams have an equal number of actions approved, which team goes first will be decided by a roll-off (that is, a representative from each team rolls the dice), with the winner of the roll-off choosing whether their team will present first or second.

Teams should present their actions in the order in which they are scheduled to begin and in the same order in which they were presented in Step 3. Continuing/ongoing actions should be presented first.

Actions are **not** assumed to be simultaneous; actions presented first are assumed to have begun first. Actions with lengthy durations are assumed to overlap other actions with lengthy durations.

NOTE: If, based on role-played S3 notes or other observation, EXCON knows that one team's actions could preempt, disrupt, or otherwise affect the outcome of one or more of the other team's actions, EXCON may impose an order or allow things to unfold based on the assigned alternation of turns. If potentially preemptive actions would logically take place earlier in a turn (they are faster, would begin concurrent with an early maneuver movement, etc.), then EXCON should arrange for that action to be presented first. If timing of a potentially preemptive action is undetermined relative to actions it might disrupt (that is, it isn't clear which action would realistically start first), then EXCON may choose to allow the actions to unfold in the order determined by the turn sequence, with benefits accruing to whichever side based on events occurring in the order in which they are presented. Guiding principles to determining EXCON decisions here are realism and fairness.

NOTE: If BLUE and RED actions are directly opposed (that is, they target the same audience seeking opposed effects, or otherwise seek effects that are directly contradictory), then those two actions must be adjacent in the sequence of presentation (EXCON may need to change order of presentation to make that occur) and will follow the separate adjudication protocol for head-to-head actions described in Sections 2.4.3.6 and 2.4.3.8.4.

NOTE: Head-to-head actions need not be the first actions presented in a turn. For the order of the head-to-head actions themselves, if one team is presenting a continuing or ongoing action begun in a previous turn and the other team is presenting a contradictory action, then the ongoing action should be presented first within the head-to-head exchange. If both actions are new but have contradictory effects, then they should be presented in the order of alternation as established above, but adjudication will be delayed until both actions have been presented and the matrix discussion will be slightly modified (as described in Section 2.4.3.6).

2.4.3.3. Presenting Team Presents

Action presentations may take no more than 5 minutes and will be controlled by a timer; any material not presented within the time limit will be excluded from consideration.

> OPTIONAL: Time limits may be waived or extended by the EXCON lead. This may include a "soft limit" whereby players are told when time expires but are allowed to finish their point.

Only the presenting player may present. That player may consult quietly with other members of their team during their 5-minute presentation period, and other team members may help with audiovisual display (advancing slides, handing around sample leaflets, adjusting pieces or icons on the game board), but only the presenting player may speak to the judges about the action.

> NOTE: Ideally, the timekeeper will have a countdown clock with a large visible display within line of sight of the players. Should this not be the case, the timekeeper will need to be more active in terms of letting presenters know when their time is drawing down, perhaps giving a warning when 2 minutes are left and again when only 1 minute is left. The timekeeper should notify players that this is coming so that they are not surprised by the warning and do not mistake it for a "you are out of time" notification.

To present an action, the presenting player should make a presentation that details the following:

- target, TA, or audience, and location
- effect desired, including the location, time, and duration of the effect
- purpose of the action and how that purpose contributes to the commander's intent and the scheme of maneuver for the phase
- forces or capabilities that will conduct or contribute to the action, where they will be located (on the game board), how they will get to that location, and when their activities will occur
- the desired end state, including what failure would look like and what success will look like
- an assessment plan, including measures of performance and measures of effectiveness, how

those measures will be collected/observed, and how long after the activities/effects measures will be reported
- three reasons the action will succeed.

In addition, the presenting player will need to present **any products that the action requires**, such as leaflets, radio scripts, storyboards for video production, and KLE talking points.

If time expires before the presentation is complete, the EXCON lead will consider whether or not there was an unavoidable interruption (such as an out-of-turn question by a distinguished visitor, or a delay due to technological problems). In the event of such an interruption, the EXCON lead may extend time as appropriate. If the EXCON lead deems that there was no such disruption, then the presentation ends immediately upon expiration of the timer.

> OPTIONAL: The presentation timekeeper may offer warnings when 2 minutes are left and again when only 1 minute is left.

During the presentation (and perhaps earlier, based on previews of actions from early intel on planned actions), the three EXCON judges should begin their scoresheet scoring. The scoresheets and scoring procedures are described below in Section 2.4.3.8.

2.4.3.4. Non-Presenting Team Rebuts

When the presenting player concludes their presentation, the timer is immediately reset, and the opposed team has 4 minutes to prepare their rebuttal, and then an additional 2 minutes to offer it. The timekeeper will track this as two separate 4-minute and 2-minute periods, making a clear announcement of which period they are in. Teams do not need to wait for the initial 4-minute discussion period to be exhausted before offering their rebuttal and may proceed immediately to presentation of rebuttal when they are read to do so; in such a case, the timekeeper should switch to counting down the 2-minute presentation period as quickly as they are able. If the non-presenting team has not completed their rebuttal when the time allotted elapses (barring adjudication of interruption or special circumstances by the

EXCON lead), they must immediately stop. Ideally, the game is played with a countdown clock visible to all players. If such a countdown clock is not available, the timekeeper may issue a timing warning at 1 minute left.

> OPTIONAL: Instead of two separate periods of 4 and 2 minutes, the non-presenting team may be given a single period of 6 minutes to discuss and present their rebuttals. This might be appropriate if the game is using a high-visibility countdown timer that shows the players their remaining time counting down.

To prepare the rebuttal, the non-presenting team may discuss among themselves, but only a single spokesperson may present no more than three reasons that the proposed action will fail or be less successful. These rebuttal reasons may take any form (a strong statement of fact drawn from the scenario materials, an assertion based on human psychology, a rhetorical question to the presenting team [which they are not allowed to answer until the counterargument]), and may address any reason that the action might be less successful: reasons related to IRC execution, design flaws in products, mistaken impressions about the TA, etc. However, **rebuttal reasons may not be actions.** This is a rebuttal, not a "counteraction," and the team must provide reasons the action will fail or be less successful because of the context, not because of something the rebutting team's forces are going to do in response. The only exception to this rule is if the rebutting team has a battle drill in place (for which they can produce documentation and which their S3 will verify) that would lead to a response action. If a team has a relevant and approved battle drill, then one of their rebuttal reasons may make reference to action that battle drill would produce.

As the non-presenting team offers their three rebuttals, the EXCON judges may take notes on or further amend their scoresheets.

2.4.3.5. Presenting Team Offers Counterarguments

The non-presenting team does not get to have the final word. After the non-presenting team completes their rebuttals, the presenting team may quickly offer counterarguments. The presenting team will have 2 minutes to confer and prepare, and then a third minute in which the presenting player (only) will present counterarguments. The timekeeper will set the clock for 2 minutes as soon as EXCON lead announces the beginning of this substep.

> OPTIONAL: The timekeeper both announces the beginning of the substep and starts the clock.

Counterarguments are limited in scope to being responses to the rebuttals from the non-presenting team. New points or issues may not be raised; only disagreements with or refutations of the rebuttal arguments are permitted. And, as is the case for the rebuttal arguments, counterarguments may not be actions. Only actions that are part of documented and approved battle drills may be presented as a counterargument. No more than three counterarguments may be offered. That could be one counterargument for each rebuttal reason, or it could be three counterarguments directed at a single particularly telling or problematic rebuttal argument.

As the presenting team offers their final counterarguments, the EXCON judges continue to take notes and adjust their scoresheets as appropriate.

2.4.3.6. EXCEPTION: Presenting Head-to-Head Actions

Sometimes actions are directly opposed by other actions, either because they seek to affect the same TA in contradictory ways or otherwise seek directly contradictory effects. These rules apply when the two teams want to present actions that, by their nature, cannot both succeed (that is, if one succeeds, the other automatically fails).

The process for presenting and rebutting head-to-head actions is slightly different. Both teams will present and rebut before adjudication. The process will begin with a standard 5-minute presentation by the first presenting team as described in Section 2.4.3.3. However, the second team will not

FIGURE 5

A Player Presents an Action in Step 4, Engagement, During IWX 20.2

Photo credit: Nate Rosenblatt, RAND.

follow the procedures in Section 2.4.3.4 (rebuttal) because they are *also* a presenting team. Instead they will follow this procedure.

After the first team presents their head-to-head action, the second team will receive a 4-minute rebuttal planning period in which they may add three rebuttals to their presentation. **The second team may not change their proposed action in any way in response to first team's presentation** (the S3 role-player for the second team should enforce this rule against any attempt to deviate by interrupting the presentation and calling attention to the foul). After no more than 4 minutes of preparation, the second team will receive a 6-minute allotment of time to present their action (as per Section 2.4.3.4 but with an additional minute allowed) and will present their presentation plus three reasons why the first team's action will fail (so they present six reasons: three in favor of their own action and three opposed to the other team's action).

After both teams have made their presentations, the first team will have a 4-minute period to prepare the rebuttal and a 2-minute period to present three reasons the second team's action will fail and offer up to three counterarguments to the rebuttals of their own action offered by second team.

The cycle concludes with the second team having 2 minutes to prepare and 1 minute to present counterarguments to the first team's rebuttals (and only their rebuttals, just as in Section 2.4.3.5).

As is the case in standard rebuttals and counterarguments, the reasons and arguments offered may not be actions or counteractions *unless* an appropriate battle drill is in place.

To review, both teams receive time to present, rebut, and counterargue in the following order:

1. First team presents (5 minutes)
2. Second team prepares rebuttal (4 minutes)
3. Second team presents and rebuts (6 minutes)
4. First team prepares rebuttals and counterarguments (4 minutes)
5. First team rebuts and counterargues (2 minutes)

31

6. Second team prepares counterarguments
   (2 minutes)
7. Second team presents counterarguments
   (1 minute).

2.4.3.7. Secret Actions

This ruleset as currently written ignores the fact that some actions taken by one side or the other would be unknown to the opposed side and would remain so, even after adjudication. Such things might include deceptions, various aspects of operations security or signature management, or even very closely targeted influence efforts. In this version of this ruleset, these "secret" actions are revealed to both teams at the time they are presented, and their effects (which would presumably also remain unknown, at least under some conditions) are also revealed.

When one team conducts an action whose result should remain unknown to the opposed side, all players should be informed of this by EXCON. Players should be advised not to act on information that they received during Step 4 but that they would not otherwise know. Players are prohibited from changing their actions to respond to or to counter actions by the other team that remain undisclosed *within the scenario*, even if they have been disclosed within the adjudication of the game. Players may need to be reminded (or re-informed) of this prohibition after the discussion and adjudication of any highly secretive action during Step 4. The reminder or admonition that "what happens in the engagement room stays in the engagement room" can be invoked to remind players not to act on knowledge about secret actions that they would not have in the scenario.

EXCON can enforce this separation between "player knowledge" (things players know from the open discussion of actions and results in matrix adjudication) and "scenario knowledge" (things that members of the OIE OPT or equivalent would know at this point in the scenario) in a number of ways. First, the maneuver aspects of the scenario follow and will remain on the paths laid out in the various established possible storylines, and these will not be changed by information that should not be known to those forces (so, even if a deception is discussed during the engagement step, unless it is adjudicated to have been revealed or failed, the opposed side's maneuver forces will continue to act as if they believe the deception). Second, the S3 role-player can push back in Step 3 on any actions that rely on knowledge that should remain unknown and decline to approve these actions.

> **NOTE:** The best way to treat secret actions was contentious in initial game design discussions, but open discussion and display of secret actions did not cause any problems in any of the actual play tests. This is likely for several reasons. First, players already understood that they were getting some information in the engagement room that they would not realistically have had in their staff position role. Second, the structure of adjudication, while allowing counterarguments, explicitly forbids counteractions, so there is no opportunity (and thus no temptation) to immediately take an action to counteract the secret action. Third, players understood that they represented a specific staff section, and that it would be the responsibility of the intelligence staff to uncover adversary deceptions and secrets and notify the commander and staff accordingly. Fourth, none of the secret actions attempted in any of the playtest wargames were ongoing or continuing actions. We recognize that the temptation to break instructions to pretend ignorance of secret actions would become stronger if the secret action was a continuous burden on a player's side, rather than a one-time/one-turn action. In such cases, a reminder by the S2 role-player in Step 1 that current friendly situational awareness does not include awareness of the secret action might help players to plan based on what their side should know rather than based on everything revealed in the engagement room.

2.4.3.8. EXCON Interrogates (Optional GREEN Input), Completes Scoresheets, Prepares for Outcome Roll

Following the presentation of the rebuttals, the EXCON judges finalize their scoresheets. Ideally, judges will have begun Step 4 with a partially and preliminarily completed scoresheet for each action based on reporting from other EXCON observers in Step 2 and Step 3 (especially from the S3 role-players) and so should be able to complete their scoring fairly quickly.

If the judges wish to do so, they may at that point make any comments and ask any clarifying questions of the presenting player or the rebutting spokesperson. Possible questions might include clarification regarding possible unintended consequences.

> **NOTE:** Both of these activities should be kept to a minimum to preserve precious time (playtesting has revealed that Step 4 tends to run long, and EXCON-related delays can be a major contributor).

> OPTIONAL: At this time, an EXCON representative for GREEN may make relevant observations about likely GREEN response or may ask questions of the presenting player or rebuttal spokesperson. These observations might also include thoughts about possible unintended effects or tertiary effects. Again, such activities should be kept short.

Each judge completes their copy of *GA8: Engagement Scoresheet*, scoring one section for the degree of difficulty of the proposed action and a second section to rank the quality of the planning for the execution of the action. These are intended to score two different things: The "Degree of Difficulty" portion of the scoresheet seeks to create a composite measure of how difficult it would be for the planned action to achieve the desired effects, whereas the "Quality of Planning" portion ranks the quality of the planning the team has done. The two scores can vary independent of one another. A team may have planned excellently to do a very hard thing or may be trying to accomplish something relatively easy based on somewhat poor planning. On both parts of the scoresheet, low numbers are preferred by the action

team: low difficulty (the effect is easy) and a low rank (first rank is best, while second rank is still pretty good, and so on).

In addition to scoring the degree of difficulty and the quality of planning for the proposed action, judges score a third section of the scoresheet—"Debate and Discussion Performance"—based on the outcome of the debate and discussion associated with the action. Judges use the debate/discussion performance modifier to add to the overall target number if the team presented and defended their action poorly in the engagement and to subtract from the overall target number if the discussion favored the proposed action.

The scoresheets support the judges in making determinations of difficulty scores of 1–10 and planning quality ranks of 1–5. The scoresheets also support the judges in making a determination of debate modifier, adding up to 2 points to an action's score if the action fares poorly in the debate and subtracting up to 2 points if the debate is favorable to the action. The scoresheets are optional, and SME judges could just make holistic determinations of the scores. The game materials include two versions of the scoresheets with different levels of structure. The first (GA8a) includes a series of structured questions with checkboxes associated with specific answers and scores. A judge completes this version of the scoresheet by checking the appropriate boxes, adding total scores, and looking up what that total score translates to in terms of a 1–10 or 1–5 ranking. The second version of the scoresheet (GA8b) is more holistic, offering criteria within a category and then asking judges to score the category on a sliding scale from 1 to 10 or 1 to 5. These sliders are then combined by eye into a final judgment rank of 1–10 or 1–5.

> **NOTE:** The presentations are required to include an assessment plan, but the quality of the assessment plan is not one of the scoresheet criteria. This is because assessment has no bearing on the success or failure of the proposed action! Assessment *does* bear on whether or not the players would know whether their action has succeeded or failed, and that is discussed in Section 2.4.3.9.2.

Each judge completes their own scoresheet. The judges then quickly confer regarding their judgments for each sheet. If they quickly reach a consensus score for each part of the sheet, then the process moves forward with those consensus scores. If they cannot reach consensus, then consensus is imposed by taking the median score for the sheet (that will be the middle score of the judges' scores). **At the head judge's option, the panel can either take the median for each component (so, a median for the different judges' 1–10 difficulty score, a median for the judges' 1–5 planning rank, and a median for the debate modifier) or they can just take the median for the final target number as determined by each judges' combined scoresheets.** The default preference for the ruleset is taking a median for each component and then combining, and instructions for doing so on *GA10: Target Number Calculation Record Sheet* are provided in the game materials.

The scoresheets will combine to produce a final target number, made up of a difficulty score that ranges from 1 to 10, a planning quality rank that ranges from 1 to 5, and a debate modifier (up to –2 for debate strongly favoring the action to +2 for particularly telling rebuttals). **Only the final target number that the presenting player must reach is announced.**

> OPTIONAL: **The median difficulty number, the median planning quality rank, and the median debate modifier are also shared aloud with both teams.** Announcing the components of the target number could give players a better understanding of why the target number is what it is, which could increase the likelihood that players will recognize the determination of target numbers as a fair process rather than an arbitrary one. Players might also benefit from knowing when target numbers are high because intended actions have been judged as difficult versus when target numbers are high because planning has been judged as poor, or because the debate has been judged as having been "won" by the other team.

2.4.3.8.1. Determining the Target Number

The target number is determined by adding the two median scoresheet scores (difficulty and planning rank) together, adding the median debate modifier, and then adding 3 to that total:[1]

$$\textbf{Target number =}$$
$$\textbf{median difficulty score}$$
$$\textbf{+ median planning rank}$$
$$\textbf{+ median debate modifier}$$
$$\textbf{+ 3}$$

This will produce a target number between 3 (difficulty 1 + planning rank 1 + debate modifier –2 + 3 = 3) and 20 (difficulty 10 + planning rank 5 + debate modifier +2 + 3 = 20).

Any outcome roll with a target number over 18 automatically fails. Target numbers that high are unlikely, because any action with really high difficulty is unlikely to be approved by the role-played S3 unless the planning execution is good (rank 1 or 2) and the presenting team makes a strong argument for the action.

In addition to announcing the target numbers (and optionally their components) to the teams, the target number calculations should be recorded on *GA10: Target Number Calculation Record Sheet*. A

---

[1] The +3 is a correction factor to drive the distribution of possible scoresheet results to align with the distribution of rolling and summing three six-sided dice. If planning rank were scaled from 1 to 10 or if there were an additional scored scoresheet component, then a correction factor might not be necessary. However, without the correction factor, the probability of an action succeeding would be much higher than is realistic. Simple, well-planned actions would be all but automatic, actions of modest difficulty that were reasonably well planned would succeed with very high frequency, and even very difficult and poorly planned actions would have unrealistically high chances of success. In addition to matching the distribution of target numbers to the distribution of possible dice rolls, this correction factor can be used for calibration of overall difficulty of target numbers. For example, in early playtests the correction factor was set at +4 or +5 instead of +3. One of the results of playtesting was a need to adjust the system of scores and target numbers to make actions slightly easier, so the correction factor was ultimately adjusted to +3.

> OPTIONAL: In a future iteration of the wargame or if the EXCON lead determines that the judges' judgments are generally too generous and actions are uniformly receiving lower target numbers than desired, the correction factor could be adjusted back up to being +4.

sample entry for the action target number calculation sheet for the RED team might look like what is depicted in Figure 6.

2.4.3.8.2. Assigning Roll Modifiers

After determining the target number for the action outcome roll, the judges may add any roll modifiers they feel they need to. Roll modifiers should be kept to a minimum to avoid pushing the envelope of the probability distribution (accidentally making a hard thing too easy, or vice versa). Modifiers should only include factors or considerations that are not included in the scoresheets or that the judges feel has not been captured correctly in the scoresheets. Reasons for modifiers might include

- accumulated prior effects from previous nearly successful actions
- complementary effects from maneuver activities
- synergies from other actions
- something important raised in the presentation or rebuttal that the judges feel is not well captured in the scoresheets.

Target number and any modifiers are announced to all players (optionally, the three components of the target number are also announced); margin for degree of success (described below) is *not* announced but should be noted.

2.4.3.8.3. Preparing the Margin of Success for the Outcome Roll

Finally, judges need to determine the margin for degree of success. This is the "horseshoes" factor for success of the action: Does "close" count at all? Is there a risk of ill effect? This ruleset offers four possible margins: *default/typical*, *partial success likely*, *high risk/high reward*, and *all or nothing*. Each is described below and summarized in Table 4.

> OPTIONAL: The narrator takes responsibility for determining margin of success and how different levels of success or failure will be described.

Obviously, the *default/typical* margin should be suitable for many actions. This includes the possibility of a partial success or a near miss, with rolls that succeed by at least 2 producing a full success and rolls that miss the target by 3 or more resulting in failure and no effect. Some actions, however, allow no middle ground: They either work or do not work and so are well suited to the *all or nothing* margin. Under this margin, any roll that equals or exceeds the

FIGURE 6
Sample Action Target Number Calculation

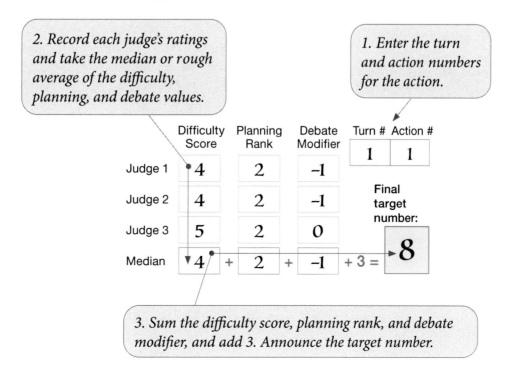

2. Record each judge's ratings and take the median or rough average of the difficulty, planning, and debate values.

1. Enter the turn and action numbers for the action.

	Difficulty Score	Planning Rank	Debate Modifier	Turn #	Action #
Judge 1	4	2	–1	1	1
Judge 2	4	2	–1		
Judge 3	5	2	0		
Median	4 +	2 +	–1	+ 3 =	8

Final target number:

3. Sum the difficulty score, planning rank, and debate modifier, and add 3. Announce the target number.

TABLE 4

Margins of Success and Their Interpretation

	Situation			
Default/ typical	Partial success likely	High risk/ high reward	All or nothing	Effect
Beat target number by				
0–1	0–3	0–1		Partial success/modest positive effect
2+	4+	2+	0+	Full success/full positive effect
5+		3+		OPTIONAL: Astounding success (see Section 2.4.3.8.3.1)
Fall short of target number by				
1–2	1–4	1		Near miss/insufficient effect
3+	5+	2+	1+	Failure/no positive effect
6+	10+	3+		OPTIONAL: Critical failure (see Section 2.4.3.8.3.1)

target number is a success, and anything that does not meet the target number is a failure. Some actions are reasonably likely to have some effect but are less likely to have the full effect. These actions might be well represented by the *partial success likely* margin, which is similar to the default margin but has a wider band for both partial success and near miss. Finally, some actions are *high risk/high reward*. These actions can have consequences for failure beyond simply failing to achieve the objective and so have the possibility of a significant negative effect as an outcome when the dice do not reach the target number (this margin might be appropriate for a risky deception, for example).

As appropriate, margin of success can be customized to suit the particular action. The four margins in Table 4 are just examples. In one form of customization, judges (or the narrator) might slightly adjust the ranges in Table 4. In another form of customization, a different margin might be chosen for success and for failure: For example, an action might not have much of a margin for a close miss but might require quite a bit more than the minimum to achieve a full success, so it might be *all or nothing* for failure but *partial success likely* for success.

Margin of success matters most when narrating the consequences of a dice roll or in cases where the

same or a similar action might be attempted next turn. For some kinds of actions, substantial bonuses (roll modifiers) should be offered for difficult actions that came close to success on previous turns.

> **NOTE:** Margins for degree of success need not be announced but should be recorded by EXCON prior to the outcome roll. There is space on the reverse of the scoresheets to note which margin of success will be used and perhaps to make notes about how the results should be described by the narrator depending on how much the dice roll exceeds or falls short of the target number; there is also an (optional) entire worksheet for narrator preparation of descriptions of each possible outcome (*GA7: Narrator's Preparation Worksheet*).

> **NOTE:** Margin of success can be adjusted based on the rebuttal. For example, if the non-presenting team makes a compelling argument that, yes, the presented action will probably succeed but that it will not be as effective as proposed, the judges might make the target number easier (or give a roll bonus) but then change the margin of success from *default/typical* to *partial success likely* to reflect that while the action is more likely to have some success, partial success is more likely than full success.

Two optional rules are to have very poor rolls indicate extraordinary failures and to have very good rolls indicate extraordinary successes. The decision to include one does not necessitate the inclusion of the other. The rules for critical failures and astounding successes are as follows:

- **Critical failures:** An outcome roll total of 3 or 4 on the dice indicates a *critical failure*: The action has failed spectacularly. As an additional option, rolls that miss the target number by a wide margin, perhaps 6 or more, can also be considered critical failures. This possibility should be determined as part of setting the margin of success for the action (see Table 4). The EXCON narrator should describe not only a complete failure to accomplish what was intended, but some sort of ill effect, too. Perhaps the action prompts exactly the opposite behavior desired from the TA, enraging them when the intent was to pacify them. Or the critical failure can be an execution failure that causes more problems for that side: Perhaps the airlift moving the IRC to the action location is lost, creating a personnel recovery challenge, failure to execute the action, and a reduction in the availability of that IRC for the rest of the game. Or a critical failure could have nothing to do with the execution of the action but could be a powerful exogenous effect (in game terms, a spontaneous inject), such as an atrocity on the part of the acting side being uncovered and publicized, or an important religious figure in another country making a huge speech that undermines the planned action. This can be especially attractive if a similar inject was planned for the next turn anyway—EXCON (the narrator) can uncork some planned misfortune and blame it on the dice.

- **Astounding successes:** An outcome roll total of 17 or 18—unless a 17 or 18 was required for success at all—indicates an *astounding success*: The best plausible result of the action occurs, assuming that everything goes as well or better than expected. As an additional option, rolls that beat the target number by a wide margin, perhaps 6 or more, can also be considered astounding successes. This possibility should

be determined as part of setting the margin of success for the action (see Table 3). Because this outcome is a consequence of dice, the explanation of events leading to this outstanding success should include elements of good fortune: synergistic effects from chance environmental factors, unexpected but coincidentally beneficial behavior from an opposed side's subordinate commanders, or some similar unpredictable but beneficial result. The execution of the critically successful action can be described as going flawlessly, too, with all assumptions validated.

> **NOTE:** For an outcome that has a target number of 17 or 18, astounding success is not possible. An action with a target number of 17 or 18 is so difficult/unlikely that *any* success is remarkable, and teams that are undertaking efforts with such limited prospects for success should not be rewarded further for luck.

> **NOTE:** Again, the inclusion of both critical failures and astounding successes is optional, and the decision to include one does not necessitate the inclusion of the other. It would be perfectly reasonable, for example, to include critical failures but not allow astounding successes. This would be consistent with the pessimistic adage of Murphy's Law: What can go wrong, will!

2.4.3.8.3.2. Preparing for Narration Based on Margin of Success

Playtesting and actual experience in IWX 20.2 revealed that the narration of outcomes after the roll of the dice is extremely important, and that advance preparation of possible narration makes delivery of the actual narrative of action outcome smoother.

The narrator should prepare notes about how they might narrate each of the possible margin of success outcomes of a roll: full success/positive effect, partial success/modest effect, near miss/insufficient effect, failure/no positive effect, and astounding success/critical failure, should those optional rules be in place. The narrator can't know in advance what the dice will deliver and so must be prepared to describe all possible outcomes, no matter how unlikely. See

Section 2.4.3.9.1 for additional guidance on narrating outcome rolls. The point here is that the narrator should prepare notes to support different possible outcomes in advance and in relation to the various possible margins of success. A worksheet for making notes related to narrating different possible outcomes is provided in *GA7: Narrator's Preparation Worksheet*.

2.4.3.8.4. Preparing for Outcome Rolls in Head-to-Head Adjudication

Head-to-head adjudication does not require a target number, so some of the procedure described above in Section 2.4.3.8 does not apply. Instead, judges will produce roll modifiers for one or both teams, with the team with the better plan, presentation, and action receiving a larger positive roll modifier. This ruleset offers three ways for judges to generate roll modifiers for head-to-head actions. Whichever method is used, the judges should take the median of the roll modifiers they individually generate and use those as the consensus roll modifiers for the roll-off.

Head-to-head actions are adjudicated with a roll-off: an opposed roll in which both presenting players will make an outcome roll, add any applicable modifiers, and then compare totals, with the highest total "winning" the opposed actions. To prevent this from being completely governed by chance, the judges will need to assign bonuses to the team whose action should be more likely to succeed and who argued better for their own position (and against their opponents').

> **NOTE:** While modifiers are discouraged in Section 2.4.3.8.2, **modifiers are encouraged here** to make the head-to-head outcome something other than a coin toss.

As noted, there are three acceptable approaches to generating roll modifiers for head-to-head actions:

- OPTION 1: Judges generate a holistic modifier for each team ranging from +1 to +6 based on any criteria they choose.
- OPTION 2: Judges award bonuses to each team based on specific criteria (such as those listed in Table 4); judges are welcome to give bonuses for

criteria not listed in Table 4, but should then be explicit about what they are giving an additional bonus for.

- OPTION 3: The downloadable game materials include *GA9: Head-to-Head Engagement Scoresheet*, which will help judges identify which team has the better action in terms of TA predisposition, magnitude of behavior change sought, effectiveness of action, planning, debate and argument, and overall, and it will help judges summarize those comparisons into a single score.

In all three options, judges should complete a scoresheet for each action in order to help them compare the relative merits of the actions. For options 1 and 2, use two copies of *GA8: Engagement Scoresheet*, one for each team; for option 3, the left side of the *GA9: Head-to-Head Engagement Scoresheet* contains an abbreviated action scoresheet for both teams.

Table 5 lists possible situations and modifiers for use in option 2 or to help shape judge assessments under option 1. None of these modifiers are mandatory, and the list is not exhaustive! If using option 2, at EXCON's option, individual bonuses can be read out to the teams, or EXCON can just read the total bonus for each side. Unless one side made a poor presentation, both sides should have some sort of bonus, but the better presentation with better prospects for success should have a larger bonus.

Modifiers should be recorded on *GA10: Target Number Calculation Record Sheet* in the special section of the record sheet for head-to-head actions. A sample target number calculation record sheet entry for a head-to-head action might look like the one depicted in Figure 7. Depending on the option used and the judges' perspectives, modifiers may be generated for one or both teams. Remember that all that really matters for the outcome is the *difference* between the modifiers (so, giving one team +4 and the other team +3 is really only a difference of +1).

> **NOTE:** Giving modifiers to both teams is a way to signal positive performance by both teams.

TABLE 5

Example Modifiers for Head-to-Head Adjudication—Option 2

Condition	Modifier/Bonus to Advantaged Team
Teams' actions/presentations relative to each other:	
Generally better presentation	+1
Point-for-point matchup of pros and cons offered	+1 per better point, max +3
Team's quality of planning rank is higher than opponents'	+2
Team's action is slightly (1 or 2) less difficult than opponents' action	+2
Team's action is significantly (3 or more) less difficult than opponents'	+4
Team's action relative to that team's other or prior actions:	
Good synergy with team's other actions	+1
Building on team's prior success	+1

FIGURE 7

Sample Target Number Calculation for a Head-to-Head Action

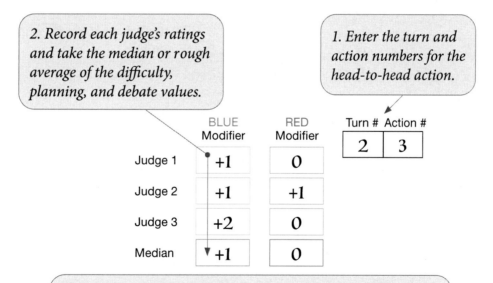

2. Record each judge's ratings and take the median or rough average of the difficulty, planning, and debate values.

1. Enter the turn and action numbers for the head-to-head action.

	BLUE Modifier	RED Modifier
Judge 1	+1	0
Judge 2	+1	+1
Judge 3	+2	0
Median	+1	0

Turn # Action #
2 | 3

3. Announce the final modifiers and which team has the advantage—BLUE, in this example—and will add the difference in modifiers—4, in this example—to their head-to-head roll.

TABLE 6

Example Margins of Success for Head-to-Head Actions

Roll Outcome		
No Effect Is Possible, and Reasonably Likely	**A Large Effect Is Possible, and Some Effect Is Likely**	**Announced Margin of Success**
Draw (dice even)	—	No effect/offsetting effects
Win by 1–3	Draw (dice even)	No effect/offsetting effects
Win by 4–6	Win by 1–2	Partial success/modest effect
Win by 7+	Win by 3–5	Full success
—	Win by 6+	OPTIONAL: Astounding success (Section 2.4.3.8.3.1)

OPTIONAL: However head-to-head modifiers are generated and whether modifiers are given to one or both teams, EXCON may choose to give more information than just the modifiers. For example, under option 2, all modifiers might be read individually and then a total modifier given. Or, under option 3, each category in the head-to-head scoresheet might be mentioned, noting which side "won" that criterion, before giving the final total bonus. While this sort of detailed list consumes more time, it would provide players better insight into why their team's action received a higher or lower overall assessment.

EXCON also needs to decide on a margin of success for the head-to-head actions. Margin of success is different in head-to-head actions because two sets of dice are rolled, so the margin must cover the possible additional variation of both rolls.

NOTE: There is no requirement that either side succeed! If it would be realistic for the tug-of-war between the two forces to result in no change in the TA's behavior, that should be reflected in the margin of success range, with one side "winning" by only a small amount having no measurable result. Or perhaps "no change" is what one side has as the intended effect and they have received bonuses because that is an easier effect to achieve, so if the other side wins the roll even slightly, there should be an effect. Table 6 provides examples.

NOTE: The narrator should prepare outcome descriptions for all possible outcomes as for standard actions as described in Sections 2.4.3.8.3 and 2.4.3.9.1. Narration for head-to-head actions include the possibility that both teams' actions are effective but offset each other, and that should be taken into account in preparing narration.

2.4.3.9. Outcome Roll, Outcome Determination

Based on the target number and roll modifiers announced by the EXCON judges,[2] the presenting player now rolls 3 dice, sums them, and adds (or subtracts) any modifiers to the dice roll result in order to generate an outcome roll. This outcome roll total is then compared to the target number (or, in the case of a head-to-head action, both presenting players roll, add their modifiers, and compare their totals to each other).[3]

The results of an outcome roll determine whether or not the action was successful and to what

---

[2] As noted in Section 2.4.3.8, the EXCON lead announces the target number and may optionally announce the components of the target number: the difficulty score, the planning quality rank, and the debate modifier. Announcing the components could give players a better understanding of why the target number is what it is, which could increase the likelihood that players will recognize the determination of target numbers as a fair process rather than an arbitrary one.

[3] Section 2.4.3.9.3 presents optional rules related to rerolls. If a team has a reroll available and their initial roll for an action falls short of the target number, they should announce their intention to use the reroll and reroll the dice.

degree (margin). **The actual consequences of that success (or failure) are determined by the description offered by the narrator, so the narration of outcomes is very important to the game.**

Whoever is assigned the role of notetaker should record the announced target numbers and modifiers, the actual outcome roll, and the key elements of the description of the outcome.

Players should also take notes on when and why their actions succeed or failed. Were they the consequences of bad dice rolls? Could they have improved their chances with better presentations or stronger rebuttals? These notes will help them adjust their plans in future turns of the game.

2.4.3.9.1. Guidance for Narration of Outcomes

The result of an outcome roll determines whether or not the action was successful and had some or all of the intended effects. **Outcome rolls determine the effectiveness of actions and would be reported by measures of effectiveness (MOEs).** Performance and execution (as reported based on measures of performance [MOPs]) are determined by EXCON narration about the outcome and follow the storyline. To repeat: The outcome roll determines whether or not the action had the desired effect and to what extent (MOEs), whereas the level of performance (MOPs) will be determined by the EXCON/narrator explanation of the why the action succeeded or failed. So, for example, if the roll indicates that the action failed (unsuccessful MOE), EXCON might decide that that occurred because of a performance (MOP) failure, such as breakdown in broadcast or printing equipment, or failure to properly load leaflets, or that it was due to some other factor in the context (for example, MOPs were good, but the audience didn't find the broadcast sufficiently credible or compelling).

While the relationship between the roll and the target number is immediately apparent (equaling or exceeding the target number is success, falling short of the target number is failure), it is still up to the narrator to make a declarative and/or narrative determination of the outcome and its consequences. These should be based on the margin for degree of success, where the roll total lands in comparison to that margin, the nature of the action, and the descrip-

tions of success and failure offered in the action presentation (EXCON should not be married to those descriptions, but they are a place to start). Possible descriptions of success or failure for each action are something that EXCON and the narrator could or should have given some consideration to during slack time in other steps so that they are prepared to describe outcomes (see Section 2.4.3.8.3.1).

> OPTIONAL: If used, astounding successes and critical failures (see Section 2.4.3.8.3.1) should be described in hyperbolic terms and should (provided the presented description was at all reasonable) meet or exceed the "what success looks like" or "what failure looks like" as described in the presentation of the action.

This verbal description of the outcome should be accompanied as appropriate by additions of icons or symbols or other adjustments to the game board.

The EXCON member acting as narrator should remember to role-play as the narrator for this story and make the description of the outcome dramatic, definitive, and entertaining. Remember that the dice do not speak for themselves, and the narrative announcement after the dice are rolled is the adjudication determination and is the culminating moment of a team's action. A big part of the excitement of the game hinges on the throw of the dice and the description of what happens by the narrator. This is how the players know that their team's actions mattered! Keep in mind that the narration needs to remain within the bounds of reality. Where feasible, narrators can refer to the presentation's "what success looks like" or "what failure looks like" outline as a guidepost. But narrators should feel free to correct these descriptions unilaterally if the presenter was too hyperbolic or too modest, as well as come up with other descriptions of the effects of the action (provided that they stay within the bounds of the overall storyline).

A narrative explanation of an action outcome should cover several things:

1. Begin with a repetition of what the target number, dice, and margin of success provided:

"The target number was 12 and the roll was 10; in this case, that is a near miss/partial failure."

2. Deliver an account of MOPs, remembering that they are completely made up as part of narration, with MOEs being what was determined by the dice.

   a. **Remember that all actions, even failed actions, have taken place or have at least been attempted.**

   b. The description of MOPs should include what the presenting force was actually able to do, even if it fell short of what was planned.

   c. Describing MOPs as falling short of what was planned can be an opportunity to soften the blow when a team has prepared well but rolled poorly—blame the failure on Murphy's Law and failures in execution: "Rough weather interfered with flight operations related to the leaflet drops; in addition, high winds affected some of the coverage of intended areas and rain reduced the readability of the leaflets as well as the frequency with which they were encountered by the target audience."

3. Describe the outcome in terms of the effect on the TA and their change in attitude or behavior. "Although some of the target audience did receive the leaflets and were moved by them, the overall impact on the target audience was insufficient to generate the effect desired. With better coverage, who knows what might have happened?" (Note that this last bit of sample narration plays on the notion of a near miss, and might encourage the team to repeat the action or amplify their influence efforts in a different way next turn). As part of describing effects, remember that the action occurred even if it didn't have the desired effects.

4. Finish with a connection to the overall operation and any impact (or lack of impact) on the operation: "The target audience continues on their natural path, which causes road congestion and other impediments to BLUE's advance on Route Bravo to continue."

> OPTIONAL: This fourth element can be held until the overall summary of impacts during Step 5 (and, held or not, it should absolutely be repeated during the Step 5 summation).

For each possible level of success or failure, the narrator should have a plan for what to report depending on dice roll result. A narrator worksheet is included in the game materials. For each possible outcome, note both the effects (in terms of MOEs and determined by the dice) and the MOPs that could have led to that. Use the following general guidelines.

For a **full success**, the narrator should use the presenter's explanation of "what success would look like" as a starting point. For a full success, performance/execution/MOPs will take place as intended and will affect the intended audience in the desired way. Preparing narration for a full success is relatively easy, as the presenters will have described what they want to happen. Narration need not be an exact read-back of what they intended to happen, however. If what the presenter requested is somewhat unrealistic, make notes for and describe a reasonable success based on the action. Also, consider additional effects (the impact on overall maneuver for both sides, for example, or secondary impact on GREEN) and be prepared to describe them.

A **partial success/modest effect** is still a success but is not everything the presentation called for. In preparation notes and in the actual narration, decide and make clear whether the shortfall was with the execution (MOPs) or with the effect (executed well, but not as effective as hoped). This can be driven by the narrator's impression (perhaps in consultation with the judges) of the quality of the planned action. If the action was well planned and relatively easy, then a partial success will have been the product of poor dice luck and perhaps should be blamed on challenges with execution (MOPs) beyond the team's control. If, on the other hand, the action was just one that was deemed as difficult/unlikely to succeed by the judges, the description could indicate good execution/MOPs but just less effect (MOEs) than desired. As to how much effect to give, that is up to the narrator's discretion. Some rules of thumb for describing partial success/modest effect:

1. Do not give the presenting team everything they asked for on a partial success.
2. Remember that the partial success is partial effectiveness—you can still narrate execution (MOP) as anything between poor and excellent.
3. If the desired effect is in any way quantifiable, give them about 50% of what they wanted (if what they wanted was reasonable; otherwise give them about 50% of what they should have expected from a full success).

For example, if the intended effect was to preclude the advance of a unit, a partial success might instead delay that unit. If the intended effect was to delay a unit, a partial success might still delay a unit but for about half the intended time. If the intended effect was to deny a unit communications in order to undermine its confidence and increase the likelihood that it remains in place, narration might indicate intermittent or patchy communications for that unit, or the unit switching to less convenient or familiar communications and so attenuation of effective communication, having some (but less) effect on morale, and delaying (but not fully interrupting) receipt of and response to new orders. Finally, where appropriate, partial successes should be described in terms that leave open the possibility of additional cumulative effect should similar actions target the same audience in subsequent turns.

Describing a **near miss/insufficient effect** can be tricky, so it is worth making notes to describe what these might look like rather than trying to come up with descriptions on the fly after the dice are rolled. Begin much as one would for a partial success by deciding how much of the blame to place on the execution/MOPs and how much to place on the effectiveness of the action.

NOTE: The decision of how much blame to place on performance versus effectiveness can have outsized impact on whether the team will try an action again. If the narration is that execution was poor, the team may propose the action again and hope for better execution. If, however, the effort is described as successfully executed but not effective, a team is less likely to try it again, *even though the dice rolls leading to those two descriptions were the same.*

In describing a near miss, be careful not to accidentally describe a partial success. If a partial success is about 50% of the intended effect, a near miss should be 10% or less. If the target number for this action was 11 or below, the main goal of a near-miss narration should be to encourage the team to try the action again (without actually saying so), hinting that the effect on the TA is heading in the right direction, they just didn't get enough of whatever was done, and better execution or continued pressure (or other actions) might produce greater effects. If the target number for this action was higher than 11, then narration might point to successful execution but limited to no effect (implying that the team should return to the drawing board with a better proposed action).

Describing a **failure/no positive effect** is easy for some actions, harder for others. Begin by deciding whether the failure is a failure of execution (MOPs) or a successful execution with failure to have intended effects (MOEs) or some combination thereof. For some actions, after describing the execution/MOPs, a simple "the action has no effect" will be sufficient. For some actions, however, taking the action has to have some effect, either positive or negative.

NOTE: Remember that **all actions, even failed actions, have taken place or have at least been attempted**. If such an action must have some effects, for the outcome **no positive effect** the narrator should describe negative effects.

Don't be afraid to shy away from really negative effects (a targeted unit having their resolve bolstered and becoming more resistant to influence or shock effects, a targeted civilian group behaving contrary to the desired behavior, etc.). Presenters' descriptions of "what failure would look like" might help, but a quick consultation with the GREEN representative may help identify other possible unanticipated negative effects. And remember that it is possible to describe the reason for failure as poor or reduced execution and suggest that the reason the action failed is that it only partially took place (which might reduce unavoidable negative effects from fully trying the action and failing).

For a critical failure, visit the full wrath of Murphy's Law on the presenting team. Either have such catastrophic execution/MOP that an asset is lost (loss of a psychological operations team, loss of an aviation asset) or a blowback of effects that causes exactly the opposite of what was desired. This may be attributable to an accident—for example, a population that was encouraged to move south actually moves north because of a translation error (if moving north would interfere with the acting side's scheme of maneuver)—or to a deception exposed and exploited, provided that the explanation doesn't overly damage the storyline. Another possibility for a remarkable failure (or success) is to impose a penalty or offer a bonus to certain kinds of actions for the next term—for example, explain that a catastrophically failed influence effort has angered and inspired the TA such that any effort targeting that same audience in the next turn will receive a roll modifier of −2.

After the action is narrated, keep track of the announced MOPs (what part of the action actually happened) and MOEs (what effects the action caused) for the summary in Step 5. This may be as easy as checking which margin of success category the final roll resulted in, if the narrator has notes for narration of all possible outcomes.

The narrator must also be mindful of the possibility that other actions presented during the turn might interfere with the level of success. Actions that might have their effect diminished later in the turn can be described as initially successful, leaving the full outcome open and to be described in Step 5 as either (1) initially successful and a successful throughout the turn or (2) initially successful but with success falling off as events develop.

For example, if BLUE's first action for a turn is a persuasive radio broadcast aimed at a segment of the civilian population, the adjudication process might result in that action being declared a success. However, if a later RED action is to jam BLUE broadcasting and it also succeeds, the narration of BLUE's success will need to change! This will be easier if the narrator is aware of the possible later action and leaves the door open to a change in narration by only describing BLUE's action as an initial success.[4]

2.4.3.9.2. OPTIONAL: Hiding Results as a Teaching Moment

As envisioned in this wargame, all players hear the target number and any modifiers announced, and then the presenting player rolls dice and generates an outcome total. Having heard the target number and modifiers, anyone who can see the dice will immediately know whether the action succeeded or failed.

However, there may be circumstances within the scenario that should prevent the results from being immediately apparent for certain actions. This might be because effects are expected to be delayed, or because the effects may not be observable until later. Or the assessment plan associated with the action

---

[4] This example would probably *not* have been flagged for head-to-head adjudication, because the TAs are different.

might be judged to be insufficient to get reliable feedback about the effectiveness of the action, either because it failed to identify good MOEs or because assets necessary to collect or measure those MOEs would not realistically be available.

If realistic circumstances would prevent players from knowing whether or not their action was effective, the EXCON lead has two choices: (1) Tell them anyway or (2) conceal the results. This ruleset prefers a hybrid course of action: combining those two options to create a teaching moment. The first time an action is taken for which available assessment is deemed insufficient to return adequate reporting, when the presenting player makes their outcome roll, the EXCON lead or the narrator (decide beforehand!) should leap forward and cover the dice after the roll but before the roll has been read (or stop the rolling player from rolling right before the roll).

Taking advantage of this dramatic interruption in the expected flow of the game, the EXCON lead or narrator should explain that, in this case, as is often true in the real world, there is insufficient information collected to reveal the results of this action. After using this teaching moment to emphasize the importance of assessment planning and the reality of fog of war in the information environment, the EXCON lead or narrator can then explain that even though in real life we would not yet know the results of this action, for purposes of the game we are going to reveal it. The dice could then be revealed and counted, and adjudication could proceed as described above.

This teaching moment can then be referred back to as a reminder every time there is an action that might not be fully observed or assessed, as encouragement to include strong assessment plans, and perhaps as a further reminder for players to plan their future actions based only on things they should know, rather than including information that is revealed in Step 4 to allow adjudication.

OPTIONAL: If there is an action taken on one turn (and which would take place during the time bounds of that turn) that should not have effects until a subsequent turn, EXCON may elect to delay the outcome roll until that later turn. In that case, the action would still be presented in the turn in which it takes place, and the judges would still determine a target number and modifiers, but the actual rolling of the dice and determination of the outcome would be held until Step 4 of the turn in which the effect should take place. In that turn, just the outcome roll portion could be inserted into the action sequence in a position deemed appropriate by EXCON, and results would be reflected in Step 5 of that turn.

2.4.3.9.3. OPTIONAL: Rerolls

Any game involving dice includes an element of chance. Dice follow a probability distribution (shown in Section 4.3, Table 9), but each individual roll of the dice is fully vulnerable to the winds of fortune. Luck represents uncertainty in the battlespace, fog, friction, chance, or Murphy's Law ("What can go wrong, will"). However, consistent bad dice luck by one team can sour what would otherwise be an engaging and entertaining exercise experience. The option for rerolls mitigates slightly against this possibility.

If rerolls will be used in the game, EXCON should decide and document the rules governing them before the game begins, including what conditions will result in the award of additional rerolls, and these rules and their intentions should be shared with players during the introduction to the game.

Available rerolls should be marked by some physical item, perhaps a poker chip or a card, that is presented to the team when they have earned a reroll and is surrendered to EXCON when the reroll is used.

A reroll is what it sounds like: a mulligan, a do-over. Some caveats apply to the use of rerolls:

1. When electing to use a reroll, the new result is final; if it is worse than the initial roll, too bad.
2. When using a reroll, all three dice are rerolled (not just one or a subset).
3. Rerolls can only be used on rolls made by the team using a reroll—that is, you cannot use

your reroll to force the other team to reroll their outcome roll.

4. Rerolls may not be used by either side in a head-to-head action (see Section 2.4.3.8.4).

If rerolls are made available, some thought must be given to how many will be made available and over what period or frequency. Rerolls might also be made available as rewards to teams who do something well.

> **NOTE:** The main purpose of the rerolls is to reduce the possibility of an unlucky distribution of rolls souring the experience of a team that is doing a good job in planning but is having bad dice luck. Emphasize ways to award rerolls that try to close the "luck gap" between the two teams.

> **NOTE:** If rerolls are given as rewards, they should reward good planning, presentation, or some other skill demonstrated by the players—not game success determined by dice. The goal of rerolls is to insulate players from some of the randomness of dice, not to reward good luck with an additional tilt toward good fortune.

Here are some possible ways to allocate rerolls (again, whatever method is chosen should be clearly stated to players so they know how and why they receive rerolls):

- At the end of each day of play, if either team has had two or more successful actions than the other team, the unlucky team gets an additional reroll.
- Each team receives two rerolls for the entire game (all six turns).
- Each team receives one reroll per game day (so, three rerolls, but only one is usable each day).
- Each team receives one reroll per turn (the authors think this is too frequent, allowing too much rerolling, but we still present the option).
- Any action that earns a median planning rank of "1" (that is, at least two out of three judges thought the action was excellently planned) receives a reroll available **for that action only**.
- Any action that earns a median planning rank of "1" (that is, at least two out of three judges thought the action was excellently planned)

earns the presenting team an additional reroll that they can use for a different period of time (that turn, that day, or any time during the game).

- Any action with a target number equal to or less than 8 (that is, any action that has a probability of success of 84% or higher) that fails (and the failure stands and is not rerolled) earns the presenting team an additional reroll that they can use at any time on a later action.
- Any other rewards-based approach to awarding extra rerolls that EXCON can think of—perhaps a reward for successfully contending with a difficult inject, or perhaps a once-per-game on-the-spot award for a particularly outside-the-box but excellent action proposal, or something else. Again, a caveat is that a reroll should not be awarded for good dice luck; rather, the opposite would be preferable—award a reroll for bad dice luck that ruins otherwise good plans.

2.4.3.10. Repeated, Recurring, or Ongoing Actions

Circumstances within the game may lead to a team repeating an action. This might be a new instance of the same action, an ongoing effort that was previously successful, or an ongoing effort that was previously unsuccessful.

The acting team will present their action regardless. However, EXCON must determine how they wish to adjudicate it. If this is a new instance of the same action, then it should be treated like a new action in all respects. This is a new attempt to execute the action. However, if the action is ongoing then alternate adjudication may be appropriate.

If an action is ongoing but has not yet been successful, then it should be subjected to the full adjudication process (presentation, matrix discussion, scoresheets, target number, outcome roll, etc.). An ongoing but unsuccessful action may be subject to one or more roll modifiers based on previous outcomes, either a bonus for cumulative effort/effect or a penalty for repeated ineffective intervention with the same audience.

If an action is ongoing and has been successful, then EXCON must decide whether to simply allow it to continue successfully or to require a new adjudication and roll. This determination can be made at the conclusion of the substep in which the non-presenting team rebuts (Section 2.4.3.4). If not much has changed since the acting team's action began and the opposed team has no new or particularly compelling reasons why the ongoing action should not continue to succeed, then EXCON may simply allow that it continues at the same level of success. This would prevent the presenting team from facing "double jeopardy" and being exposed to multiple dice rolls (and chances of failure) for something that has already been determined to be successful.

If, however, circumstances have changed that would in any way call the continued success of the action into doubt, then it should be re-adjudicated and a new outcome roll should be made. The judges may choose to give a roll bonus as appropriate for continuing a successful ongoing action even if circumstances have changed such that it is no longer a slam dunk. If an ongoing action is now subject to direct counteraction by the opposing team, then it should follow the procedures for head-to-head adjudication as described in Sections 2.4.3.6 and 2.4.3.8.4.

NOTE: Requiring teams to present their ongoing actions for approval and for adjudication helps maintain a check on resources. The EXCON members role-playing S3s may quickly re-approve continuing actions, but will consider that ongoing action and its resource requirements against the overall resource draw of the team for the turn, which may help constrain the overall total number of actions.

NOTE: If an action to be taken by one team has the potential to interfere with the ongoing action of another team, the potentially interfering action should be presented and adjudicated first. The ongoing action (carrying over from an earlier turn) can be assumed to continue succeeding until the time of execution of the interfering action.

The ongoing action can be re-adjudicated after the interfering action, with modifiers as necessary based on the success or partial success of the interfering action. If the level of possible interference of the new action meets the criteria for head-to-head adjudica-

tion, then follow the steps in Section 2.4.3.8.4, with the continuing action presented first and the interfering action presented second, with both actions adjudicated simultaneously.

2.4.3.11. Repeat Activities as Necessary

The substeps described from Sections 2.4.3.3 through 2.4.3.10 should be repeated for each action until all approved actions for the turn have been presented and resolved.

2.4.3.12. Record All Results and Prepare for Step 5

EXCON members should have been taking notes on everything throughout, but as the process moves into Step 5, it is particularly important to make doubly sure that outcomes from all actions have been recorded, as have the final positions and dispositions of elements on the game board.

2.4.4. Outputs from Step 4

Step 4 should produce notes on the actions completed, their success or failure, and their outcomes as described by the narrator. Final dispositions of actions as adjudicated on the game board should be recorded and left in position to allow a cohesive narrative in Step 5. Output for filing and recording should include all of the completed scoresheets for all of the actions.

2.4.5. Time Allowed for Step 4

Step 4 has the greatest number of activities and outputs of any step within the wargame, and, even though the procedure is fairly structured, it has the potential to take longer than anticipated. If Step 4 is run smoothly and efficiently and teams are limited to two actions each, it might be concluded in as little as 60 minutes. With many actions to adjudicate or with actions that stress the processing and scoring capacity of EXCON, Step 4 might stretch to 90 minutes or beyond. Initial target time allocation recommendation is to allow 90 minutes for Step 4; if it runs over, Step 5 can serve as a buffer, and the exhaustion of the allocated 90 minutes can serve as a signal to the member of EXCON running the engagement meeting to accelerate and conclude the proceedings.

2.5. Detailed Conduct of Step 5: Results and Reset

2.5.1. Overview of Step 5

Step 5 should begin immediately on conclusion of the resolution of the final action taken during Step 4, and should occur while both teams are still in the engagement room. The players get to relax for a few minutes while EXCON completes Step 5, recording all outcomes, narrating the results and consequences of all effects, and determining their impact on the overall progress of the scenario.

2.5.2. Inputs to Step 5

Most inputs should already be present: projection capabilities, game board, results of actions, and dispositions of actions represented on the game board. Additionally, the anticipated results for maneuver for the turn should be ready and ready to present.

If this is the final turn, inputs include the preparation of any awards that will be presented.

2.5.3. Player Activities During Step 5

Players should be attentive and take notes. Players also should get fired up: either to follow up their successes and stick it to the other side, or to lick their wounds and get back on their feet. The next turn is a new day! In the final turn, players should stand by with bated breath waiting for the final adjudication of the wargame and the presentation of awards.

If players have an urgent question or if something in the storyline is unclear, they may bring it to the attention of EXCON during Step 5. Such questions should either be addressed to the EXCON lead or the narrator as appropriate.

2.5.4. EXCON Activities During Step 5

EXCON members should quickly meet and determine whether any of the actions for the turn have any consequences for the projected trajectory of maneuver outcomes and the storyline for the turn and, if so,

quickly identify these changes. Such changes should not be a surprise, and adjusting for them should be quick and easy. Anticipating such changes should be the responsibility of the narrator, though the narrator may rely on other members of EXCON for support.

The narrator then presents a summary narrative of the turn, noting the outcomes of all actions from Step 4, explaining the immediate impact of those actions on maneuver, and summarizing the overall maneuver outcomes/progress for the turn. **The core goal of this summary is to take the results of OIE actions from the teams for the turn and put them back into the context of the broader operation.**

The narrator should take pains to make it *sound* like successful and potent actions are having an effect on the context and on maneuver, even if maneuver plans remain within tolerance and are pretty much on-script. In the wargame, as in real life, perception is reality. If players believe that their actions have affected the shape of the operation, then they will feel much more empowered and consequential. Again, the narrator should adopt a role-play attitude and should describe both the OIE and maneuver and other operational progress dramatically.

The presentation in Step 5 is a summary of everything that has happened (OIE and maneuver) during the turn. This is different from the Step 1 presentation, which is a summary of the current state at that time and projection of what is scheduled to happen during the turn to come.

After concluding the summary of actions completed and providing the update to the storyline covering what has happened during the turn, EXCON should caution players about which of the results and consequences from the turn's actions are "scenario knowledge"—things the OIE OPT would be aware of in real life—and which are "player knowledge"—things that the players got to see adjudicated with the God's-eye view of the engagement in front of EXCON, but that in the context of the scenario remains shrouded by the fog of war and the limitations of assessment. The phrase "what happens in the engagement room stays in the engagement room" might be a good reminder of this distinction.

Members of EXCON should also be preparing for Step 1 of the next turn. Materials for Step 1 of the next turn should already be pre-prepared and follow a pre-prepared storyline but may require adjustment based on the rolls and outcomes of the current turn. This may need to happen simultaneously with the summary of the turn's results by the narrator, and so it may need to involve other personnel (perhaps in collaboration with those who will be presenting the Step 1 update). This reset huddle typically takes 15–20 minutes to complete.

Step 5 requires one person from EXCON (the narrator) to give the summary narrative/storyline for the turn. It also requires a few (two or three) EXCON personnel to quickly consult with each other and adjust the storyline for maneuver based on OIE action results. Additionally, the same EXCON personnel who provided IT and game board support in the previous step should continue to do so.

Finally, two to four EXCON personnel (one or two for each team) should be listening to the narrative and changes and preparing any adjustments necessary for the update presentation in Step 1 of the next turn. Five to nine EXCON personnel are required for this step, depending on how many of the roles are included and which, if any, are either dual-hat responsibilities or are assigned more than one EXCON representative.

If this is the final turn of the game, the narrator will not only summarize the turn but will also narrate the conclusion of the conflict, indicating the winning side and the winning team (which need not be from the winning side). EXCON will present any awards at this time. (See Section 2.5.5.1.)

2.5.5. Outputs from Step 5

Step 5 produces an updated situation to inform the next turn, including possibly durable changes to the game board and adjustment of the expected track for maneuver. Step 5 also produces notes and records for after-action review and subsequent turns. In Step 5 of the final turn, outputs include various awards (ideally accompanied by certificates, but certainly by good feelings and good cheer; see Section 2.5.5.1).

2.5.5.1. Outputs in the Final Turn of the Game

The game will end on a Step 5 of the final turn. At that time, the narrative summary of the turn should be expanded to cover not only the turn, but the trajectory and course of the whole operation, as well as the outcome. A winning team should be identified based on the extent to which they were able to support their side's concept of maneuver and impact the overall trajectory of operations.

NOTE: The winning team need not be from the winning side. One side in the conflict may be much stronger and destined to prevail in the conflict, but excellent performance by the weaker side's team might make that victory take longer, or be less complete, and that should be acknowledged and rewarded. For example, even if the BLUE operation succeeds, the RED team might have made it so much more difficult on the BLUE force that they "won" in the information environment.

Ideally, all members of the winning team should receive a winning certificate. Other awards might also be presented, including humorous ones. Awards can be to individuals or for actions. Awards for actions will go to the presenting player for that action, but with recognition that others from the team will have contributed to the development of that action.

Examples of possible awards include best presentation, best rebuttal, best presentation of a failed action, action with biggest impact on maneuver, action most emblematic of the principles of OIE, EXCON's favorite action, most embarrassing proposed action, action furthest outside the box, least likely to be friends with the Judge Advocate General, and "better lucky than good" (for an action with a high target number that succeeded anyway).

> OPTIONAL: The final narration, declaration of the winner, and the presentation of the awards may be held until the wargame after-action review.

2.5.6. Time Allowed for Step 5

Step 5 should usually not take very long to execute, as results should fit within one of the predetermined tracks (storylines) for overall operational progress, and many of the effects and results will have been previewed during the adjudication of actions throughout Step 4 and anticipated as possibilities by EXCON based on their awareness of what was being presented in Step 3. However, if outcome rolls have been extreme or if EXCON has been surprised by a team's actions, more time may be required.

Initial recommended allocation of time for this step is 30 minutes, with the understanding that it may be used as buffer for an overflow from Step 4, or may transition directly into Step 1 of the next turn.

> OPTIONAL: In Step 5 of the last turn of the day (except on the last day of the wargame), players may be excused after information relevant to them has been shared, while EXCON continues to do bookkeeping in preparation for the next turn's Step 1.

> OPTIONAL: Step 5 of the last turn of the day may immediately segue into Step 1 of the first turn for the following day, to allow players to make their Step 2 preparations overnight.

2.6. Troubleshooting Problems During the Game

It is not unlikely that problems of some sort will arise during the game. Players may take an action that surprises EXCON or a role-player, or an extreme value on an outcome roll may dictate a surprising catastrophic failure or astounding success that really should have greater impact on the scenario than what has been prepared for in the existing storyline options. When such a situation occurs, keep calm and carry on.

Two principles from other contexts offer guidance. First, "the show must go on." This wargame is not entirely unlike a live theater event, and so the game must continue without major interruption, even if that requires generous ad lib from EXCON. EXCON personnel should put their heads together,

quickly talk it through, then decide on how to proceed. Second, "when in doubt, roll and shout." If something cannot be reasonably and fairly adjudicated (it is outside the scope of the scoresheets and parameters established by this ruleset, or should be treated as genuinely contingent, or an arbitrary ruling would unfairly favor one team over the other), then the EXCON lead (or the narrator) should declare a target number, make an outcome roll, dramatically announce the meaning of the roll, and proceed with that outcome as the final adjudication of the matter.

2.6.1. Dealing with Turn Outcomes Getting Out of Bounds

As envisioned (and described in Section 3.4.5), the wargame has a range of possible outcomes for each turn, driven primarily by predetermined maneuver outcomes but capable of being affected by OIE along three tracks or storylines: (1) an expected direction of progress, (2) progress based on RED OIE outperforming, and (3) progress based on BLUE OIE outperforming. If teams and dice are relatively balanced, turn-by-turn progress of the overall operation should remain within the bounds envisioned by the scenario. However, they might not. If progress threatens to get out of bounds, there are two tools to address it, described in the next two sections: skipping or repeating a turn and using corrective injects.

2.6.1.1. Skipping or Repeating a Turn

One way to get the game back in bounds might be to skip a turn or repeat a turn in the planned sequence. If the planned bounds for the current turn will not accommodate what should be happening based on the outcomes of the actions, but the outcomes for a future turn will, consider skipping to that turn. Or, if the amount of progress implied by even the lower bound is not plausible given the disruptions caused by OIE, repeat the turn, asserting that the attacking force basically made no progress and needs to keep trying before they can progress to the next phase.

These notional situations assume that one force is the aggressor and is looking to skip turns/phases and get ahead, while one force is the defender and is looking to slow (or completely derail) the progress of the other side. Using either turn skipping or turn repeating might force a change in the total turn count. This would either reduce the game length (aggressor wins early) or require an extra turn (defender significantly delays the inevitable or perhaps even prevails). The EXCON lead will ultimately make the call.

> **NOTE:** The idea of skipping or repeating a turn could be interpreted in at least two ways, and both could be valid in this wargame. The first underlays the intent of the preceding paragraph and just means that the operation does not progress along the storyline as intended because the actions of the (probably defending) force have prevented the operation from moving forward to or toward the next phase as scripted and the teams must fight another turn under pretty much the same scenario circumstances. In that case it would still be a new turn, but the Step 1 update of conditions would be virtually identical to the Step 1 update from the previous turn.
>
> The second possible interpretation is a "do-over" or "mulligan" where teams are instructed to forget what happened in the turn and start over and try again. This latter interpretation should be avoided if possible, but might be necessary if one or both teams fail to effectively execute most of the steps of a turn and everyone really just needs to try again.

2.6.1.2. Intentionally Causing Problems During the Game: Scenario Injects

It wouldn't be a wargame without the possibility of surprises, which in wargames are termed *injects*. We envision the possibility of three kinds of injects: *variation injects*, *challenge injects*, and *corrective injects*.

Variation injects (or "white noise" injects) simply stimulate the teams to be dynamic and responsive. By presenting an unforeseen event (or just chatter about a routine event or even non-event), the team might have to adjust its plans and actions to take account of the change in context. Use of variation injects should only be necessary if the opposing teams are not providing much unanticipated stimulation, either because their actions aren't surprising or interesting, or because their actions aren't succeeding (bad dice luck).

Challenge injects make the game harder for one or both sides. If things are going too smoothly, challenge injects don't just provide something unexpected to stimulate the players—they provide something *difficult* to stimulate the players. This could be something like a significant civilian casualties incident. Challenge events are generally one-sided, in that while they might require a change in plans by both sides, they are only bad/challenging for one side. To balance this, the total number of challenge injects planned for the game should be even for both sides, even if they don't occur on the same turns (for example, RED could face challenge injects on turns 2 and 4, and BLUE could face challenge injects on turns 1 and 4). Challenge injects should be balanced, unless they are being used as corrective injects.

Corrective injects are only used if the need arises. A corrective inject adjusts for one team totally outperforming the other team, or the dice brutally disadvantaging one team but not the other. The intent of the wargame is for it to be fair and for success or failure to be real possibilities depending on play, but too astounding of a degree of success by one team too early in the game could push the pre-scripted material out of bounds. One way to get the game back in bounds might be with a corrective inject, which is basically a challenge inject that is not paired with a balancing challenge inject for the other side.

Injects of any kind should be announced during Step 1 of the turn on which they occur. As appropriate, both teams should receive information about the inject based on what that side would know.

Injects should be prepared on inject cards, which can then be given to one or both teams as they prepare their turn (note that some inject cards will require a RED and a BLUE version of the card, as events may be perceived differently be the different sides). Ideally, the scenario should include a whole deck of inject cards of each type in order to allow EXCON some preplanned flexibility. This ruleset does *not* advocate random draws from the inject deck; the deck format is just to provide variety in what is available, not to add an unnecessary (and potentially risky, from the EXCON perspective) random element. See Section 3.4.6 for more on preparing scenario injects.

2.6.2. Dealing with Uncertainty About How Things Actually Work

The U.S. armed services have a wide range of platforms and capabilities; if partners and adversaries are included, the world's military capabilities are many and varied indeed. It is unrealistic to expect EXCON to know everything about the operation and performance of all of them. While someone on EXCON should have some knowledge of all of the core capabilities assigned to the two sides (see Section 3.4), that knowledge cannot be exhaustive. At some point in a game, a question about the operation or effect of a piece of gear, a platform, or capability will come up that no one on EXCON is immediately prepared to answer. Such uncertainty might pertain to a process or a regulation, too.

How to respond to urgent uncertainty depends on when it comes up. If a question about a capability comes up during Step 1, 2, or 3, saying, "I don't know but I'll get back to you" is perfectly reasonable, provided that someone from EXCON can then attempt to run the needed facts to ground. However, if the uncertainty comes up during Step 4 (engagement), a response may be needed in order to adjudicate the current action, and a delay for research may not be possible. Moreover, such uncertainties are reasonably likely during engagement, when one side has been making an assumption about the operational parameters of a capability that the other side (rightly) calls into question during their rebuttal.

In such a situation, where there is uncertainty and urgency, the EXCON lead should make an adjudication, either making an expert judgment or declaring it a 50/50 toss-up and making an unmodified outcome roll, in which a roll of 11 or higher favors one side's interpretation and a roll of 10 or less favors the other. Regardless of how the uncertainty is resolved, every effort should be made to find a source (or a SME) that can provide the correct answer, and that answer should be shared with all participants for education purposes. Note that whatever was decided at the time is how the capability worked in the scenario universe during the relevant turn and action. Even if the wrong adjudication was made, the story will not be rewritten to accommodate what the adjudication should have been. The past is the past, and the game only moves forward.

2.6.3. Dealing with Players Who Game the Game or Stress the Bounds of Civility

Despite guidance to the contrary in the *Player's Guide* and in the introductory presentations to the players, some players' competitive drives can lead them to behavior that runs counter to what is ideal for the success of the wargame as a training event. For example, although we did not encounter this challenge in any of the playtests of this wargame, in other wargame contexts we have seen participants "fight the scenario"—that is, engage in extended disagreements about what is fair or realistic—or "game the game"—that is, propose actions or moves that would not make sense for someone in their assigned role in the game scenario but might give them or their side an advantage within the structure of the game. One challenge we did see in one of the playtests involved a player becoming hostile and argumentative toward members of the opposing team and stretching the bounds of civility during Step 4.

One of the keys to avoiding challenges like this in the first place is in the statement of expectations at the outset of the game, and in the social climate established throughout the exercise and the wargame. EXCON and other leaders can display attitudes and lead through examples of the "right way" to play and to engage with the game and other participants. Should that prove insufficient, the Marine Corps has a strong and healthy leadership tradition of taking errant subordinates aside and counseling them regarding their behavior. This proved effective with the player whose exuberance led him to cross the bounds of civility; after being pulled aside by a leader, the marine remembered himself and the context, apologized to the opposing team, and continued to compete, debate, and disagree with them, only in an appropriately respectful way.

3. PREPARING FOR THE IWX WARGAME

This section contains notes, observations, and suggestions related to preparing the wargame prior to execution. Many of these issues and activities should be considered as part of the planning conferences and other preparation efforts for the IWX of which the wargame will be a part.

3.1. Selection and Assignment of the Players

As briefly noted in Section 1.5, this game is designed for teams of 6–10 players. Playtesting has revealed that as few as 3 engaged and experienced players (experienced with OIE and planning, not necessarily with wargames or with this wargame) can form a sufficient team. Teams larger than 10 should be avoided, as they could start to become unwieldly and hard to manage (both from a team leader perspective and from an EXCON/training perspective).

Available IWX participants can be divided into players on the two teams by any expedient manner. Players/teams can be assigned by exercise staff, or exercise staff can choose two team leaders and allow them to choose teams (perhaps using some variation of "schoolyard pick 'em," or something more akin to an NFL draft).

Allowing team leaders to pick teams is certainly a fair approach to begin a competitive game, but it may not be optimal for targeted training and education purposes. The following are some considerations that might cause EXCON selection of teams to be preferred and criteria that might be used:

1. There might be certain personnel that will deploy soon in a certain role, and having them practice that role during IWX would be to their benefit (for example, making sure they are on the BLUE team and have a certain role within that team, such as team leader).
2. Ideally, each team will have a distribution of capability backgrounds and expertise; if there are two public affairs specialists and three psychological operations specialists among the participants, each team should have at least one of each. That might happen with participant-picked teams, but it can be assured with EXCON-assigned teams.

3. Ideally, levels of seniority and experience will be divided between teams. A situation where all the senior personnel with practical experience are on one team and the other team is composed of exclusively more junior personnel might both make the wargame less fair and limit opportunities for within-team mentorship.

3.2. Requirements for EXCON Personnel

The wargame requires at least 5 EXCON personnel to execute. Optional additional support could engage 5 or more additional personnel—see Table 7 in Section 3.3. These individuals should meet these general requirements:

- Be available to participate in the full span of IWX and available for EXCON rehearsals/training (as detailed in Section 3.2.5).
- Have some experience with operations in the information environment, their planning, and the various information-related capabilities. IWX and OIE working groups, simulated or otherwise, tend to have acronyms flying thick and fast, and EXCON personnel must be comfortable with the vocabulary and the flow of such efforts.
- Have some experience or at least familiarity with OIE training events or exercises like IWX and some familiarity with wargames (either as a hobby gamer or though observing or participating in defense wargames).
- Be patient, detail-oriented, and willing to read and reread rules.
- Be able to work collaboratively with others.
- Be able to stay within their assigned roles and not default to teaching and instruction (unless their role during the game explicitly involves teaching and instruction).

In addition, some of the specialty roles require additional specific skills, as described in Sections 3.2.1–3.2.4.

3.2.1. Requirements for the Operations Officer (S3) for Each Side

The role-players for the operations officers/S3s (or RED equivalent) are absolutely critical. These must be veteran and expert personnel who are capable of challenging the players, forcing them to defend, explain, and justify their proposed actions, and giving them a realistic "OPT lead briefing the S3" experience. In addition to the required experience and force of personality, these role-players must also be very well versed in the details of the scenario and the operation.

The S3 role-players are responsible not only for giving the players an appropriately hard time, but also for ensuring that the correct number of well-planned and -presented actions arrive in Step 4.

It is imperative that the same role-players represent these critical roles throughout the entire game, not just for continuity, but for consistency in the filtered view of the scenario and operation that is provided.

It is also imperative that the S3 role-players protect their role as hard-nosed S3s. As such, they should avoid too much hallway jocularity and should ideally not be dual-hatted as a senior mentor. Their role is the boss and the evaluator, not the coach.

A list of S3 role-player duties by game step is provided in *GA5: S3 Role-Player's Responsibilities*.

3.2.2. Requirements for Intel Chief (S2) for Each Side

The S2 role is less about role-playing (the S3 is *all* about role-playing!) and more about intimate familiarity with the scenario and the forces of which the RED and BLUE teams are a part. The S2 role-player need not have an intelligence or wargaming background but must be able to confidently and comfortably give an update briefing.

If the S2s are also the map managers, they must be comfortable working with maps and tokens or pieces and familiar with the map iconography and symbology used within the game.

Personnel involved in scenario development would be a good fit for the S2 role, as they will already have much of the relevant knowledge and may even have been involved in developing the materials.

A list of S2 role-player duties by game step is provided in *GA4: S2 Role-Player's Responsibilities*.

3.2.3. Requirements for the Judges

Judges may be either civilian or military and should be relatively senior. Judges should be well-versed in the Marine Corps Planning Process or the Joint Operation Planning Process.[5] Judges must have experience bordering on expertise in OIE, information operations, and one or more IRCs. Judges must have discriminating judgment and the ability to discern what is practical from what is impractical and what is easy to accomplish in the information environment from what is hard.

Judges must be familiar with the scenario, since team actions must be evaluated in context. A panel of judges who arrive at scores by consensus is preferable to a single judge, and a panel of three judges is recommended. Judges must be familiar with the game's scoresheet (or willing to gain familiarity through EXCON training/rehearsals, as described in Section 3.2.5).

A list of judge responsibilities by game step is provided in *GA3: Judge's Responsibilities*.

3.2.4. Requirements for the Narrator

The narrator should have a good baseline of expertise related to OIE, a high level of mastery of the game rules, and close familiarity with the scenario. However, these things alone are not sufficient to fulfill this role. The narrator needs to have a certain level of something like the "gift of gab" and a certain degree of quick-thinking creativity. Individuals with the right stuff for narration might well have a background that includes improvisational theater, might have experience as a "dungeon master" or "game master" for tabletop role-playing games, or could just be an accomplished and captivating storyteller. These

---

[5] Marine Corps Warfighting Publication (MCWP) 5-10, *Marine Corps Planning Process*, Washington, D.C.: U.S. Marine Corps, as amended May 2, 2016; and Joint Publication 3-0, *Joint Operations*, Washington, D.C.: U.S. Joint Chiefs of Staff, incorporating change 1, October 22, 2018.

are not the only ways to qualify, but note that the needed qualities should not be assumed to be universal among otherwise intelligent and accomplished professionals. There is an art to this, and not everyone has the necessary talent.

Playtesting confirmed that the narrator role is both critical and challenging. It is essential that the right person be found for this role, and that it not be considered a "prestige" position offered to the most prominent EXCON representative, for example.

A worksheet to help the narrator during Step 4 is provided in *GA7: Narrator's Preparation Worksheet*.

3.2.5. Training and Rehearsals for EXCON Personnel

Because this wargame is not exactly like any other wargame, EXCON personnel will need a certain amount of exposure to and practice with this ruleset before they will be able to run a smooth game. Playtesting definitely revealed benefit from familiarity with the flow of the game, experience with the scoresheets, etc. New members of playtest EXCON personnel frequently noted that there is a bit of a learning curve!

All EXCON personnel should carefully read this entire ruleset (and the associated *Player's Guide*) as part of their preparation for the wargame. Even personnel who have participated in a previous iteration of this wargame are encouraged to reread the rules. The rules cover a wide range of circumstances and contingencies and so are fairly dense, especially for a first-time reader. Even a careful read is not wholly sufficient to prepare EXCON personnel. EXCON should also schedule one or more rehearsals, dry runs, walkthroughs, or "Map EX"–type preplanning sessions to ensure that everything goes smoothly. Such rehearsals might include any or all of the following topics:

- Introduction to the purpose of the game
- Review of the vocabulary and terms of the game (see Section 1.3)

- Introduction to the artifacts of the game: the game boards and pieces, pointers, timers, the scoresheets, the dice
- Assignments of EXCON roles for each step of each turn (see Section 3.3)
- Discussion of the responsibilities in each step for each role
- Guidance for S2 and S3 role-players (and any other role-players being used)
- Practice for the S2s giving the Step 1 update briefing
- Guidance for the narrator
- Practice for the narrator in describing different outcomes (success, partial success, near miss, failure) for an action
- Guidance for the judges
- Practice for the judges in using the scoresheets to score actions
- Discussion between the judges about the different criteria and the judging standards
- Step-by-step pacing and spacing of the game: who needs to be where, and when
- Layouts of the rooms: where different EXCON personnel will sit, especially for Step 4
- Number of game turns planned, phase of the operation to be covered by those turns, number of actions to be allowed per turn
- Briefings related to the scenario, including the core "storyline" for maneuver
- A chance to ask questions and have them answered.

Members of EXCON should all also be fully familiar with the exercise scenario, both whatever materials have been or will be provided to players as background on the operation and whatever materials have been prepared as part of the storylines for maneuver and outcomes as the game progresses (see Section 3.4.5).

3.3. EXCON Personnel and Roles, by Step

Table 7 matches activities within each step (the rows) with EXCON individuals and their overall role (the columns), listing minimum core personnel and where additional EXCON personnel could be assigned roles. All assignments of roles to specific individuals are provisional and may need to be tailored to the specific personnel and circumstances of the actual game. Table 7 includes some distinctive notation:

- "X" indicates that the activity will normally be performed by the EXCON member listed at the top of the column, if there are enough EXCON personnel for that role to be assigned.
- An asterisk (*) indicates that an activity is suitable for assignment to another EXCON member if necessary or practical.
- "A" (for "alternate") indicates that the activity is suitable for assignment to the EXCON member listed at the top of the column instead of or in addition to the member indicated by the "X" in the row.

If the S2 and S3 role-players are going to be used for an EXCON role outside the control room other than as an S2 or S3, then these role-players should have a neck placard indicating when they are "on" in their S2/S3 role versus serving as just another member of EXCON. Ideally, and personnel permitting, the role-players should only participate as the S3 in Step 2 (or the S2 in Step 1) and in "control room" activities during other steps, unless they are required to play additional roles. There is no problem with S2 or S3 role-players being part of the EXCON team for Step 4, as Step 4 is pretty much all hands on deck; if an S2 or S3 is used as part of the judging panel, then S2s or S3s from *both* teams should be judges, as it might be perceived as unfair to have one side's S2 or S3 as a judge without the other to balance things out.

OPTIONAL: As a further optional role, EXCON might include someone explicitly responsible for representing GREEN. GREEN status is typically maintained as part of the scenario and as a general EXCON duty, and GREEN status updates are typically provided by the S2 role-players (or whomever provides the Step 1 updates), but a separate EXCON role could be assigned for these responsibilities. If there is an EXCON GREEN representative, that individual could also be used on the panel of judges for Step 4.

OPTIONAL: Playtesting revealed a need to have someone with control over and the final word on reality within the scenario: what works, what doesn't, and the status and condition of things not spelled out explicitly in the scenario. We call this role "reality master," but it could also be called "king of the world" or "scenario control." This role could be filled by the EXCON lead, but playtesting also revealed a relatively heavy burden on the EXCON lead from existing administrative and other duties within the game. This role could instead be filled by the lead of the team that prepared the scenario, by the narrator, or by a member of the narration support team. Optionally, this role could be combined with responsibility for game board management. The reality master would explicitly have the following responsibilities:

- Be the final arbiter of any RFIs.
- Be the final arbiter of capabilities available at any time for each team and for facts about their capability and functioning.
- Track all answers given and rulings related to the above points to ensure continuity and consistency in the scenario world.

TABLE 7
EXCON Personnel, by Role and Activity in Each Step

Activity	Required Personnel					Optional Roles for Additional Personnel					
	EXCON Lead	BLUE S3	RED S3	BLUE S2	RED S2	Note-takers‡	Narrator	Reality Master	Time-keeper	GREEN Rep	Additional Judges‡
Step 1											
Update BLUE				X							
Update RED					X						
Notetaker to BLUE						X					
Notetaker to RED						X					
GREEN brief (optional)				A	A					X*	
Step 2											
EXCON representative(s) available to answer questions				A	A		A	X*			
Observer/notetaker for BLUE						X					
Observer/notetaker for RED						X					
Step 3											
S3 role-player for BLUE		X									
S3 role-player for RED			X								
Notetaker to BLUE						X					
Notetaker to RED						X					
Step 4											
Judging	X	A	A								X
Timekeeping				A	A	A			X*		
Game board management				A	A	A		X*			
Notetaking						X					
Narration of action outcomes	A						X				
GREEN brief (optional)	A									X*	
Step 5											
Adjust outcomes/storyline from results (team effort)	A						X				
Deliver turn outcome summary	A						X	A			
Game board management				A	A	A		X*			
Consolidate changes to storyline for step 1 of next turn (team effort)	A						X				

"X" indicates that the activity will normally be performed by the EXCON member listed at the top of the column, if there are enough EXCON personnel for that role to be assigned.

* indicates that an activity is suitable for assignment to another EXCON member if necessary or practical.

"A" (for "alternate") indicates that the activity is suitable for assignment to the EXCON member listed at the top of the column instead of or in addition to the member indicated by the "X" in the row.

‡ If notetakers and/or extra judges are available, two of each are recommended.

3.4. Preparing the Scenario

This ruleset is designed to be scenario-agnostic, provided that the scenario pits a notional BLUE force against a notional RED force in an operational context where efforts can be supported with OIE. But the scenario itself requires a great deal of detail and preparation. This section provides guidelines and tips for developing a scenario and providing a sufficient amount of detail within that scenario to meet the requirements of this wargame.

The scenario needs to include sufficient detail to allow teams to plan OIE to support their side's operation. This should include information about context, with a focus on things relevant to OIE. The scenario should include details on things that may not be included in standard intelligence preparation of the operating environment, such as information on the intentions of various noncombatant groups, the different communications infrastructure present and use patterns, important leaders within the two forces, the personalities and proclivities of those leaders, and characteristics related to will to fight for both RED and BLUE forces.

Initial briefings should only include information realistically available to the two teams in their respective roles, but other information related to the information environment that might be uncovered through RFIs or through experience should be included within the full scenario details so that EXCON is not called on to invent such details on the fly during the game. The best way to ensure that a scenario is sufficiently comprehensive to answer a variety of questions that may arise is through playtesting.

Remember that this is a two-sided game, and full intelligence preparation of the operating environment materials must be provided to both sides.

Scenario information provided to each side must also include lists of the various maneuver elements involved in their force, the IRCs that are organic to their force, and IRCs that might be available by request from a different echelon or adjacent force. Each side should receive that same information about the opposed force, filtered or obscured as appropriate because of realistic intelligence constraints.

In many wargames, the players or participants are the source of all plans for the formation they represent, and the scenario only supplies the starting position and context. This is not the case for scenarios for this wargame. Because the players only contribute the plans and action related to OIE, all other plans and activities by their notional side must be part of the scenario. So, the scenario is not just the starting position and context but is also the story of exactly how the whole operation will unfold absent the input of the teams' OIE.

> **NOTE:** The results of the OIE may cause EXCON to need to slightly adjust that story, but the story of the operation written as part of scenario development will pretty much stand (see Section 3.4.5). One EXCON playtester described the situation like this: "Imagine that the operation, the game, is over. Now write the history of everything that happened except the OIE part. That is what we need to know for the scenario."

Obviously, only the starting position, context, and initial maneuver plans for their side are provided to players. But the planned evolution of the context and the plans for both sides, as well as the expected maneuver outcomes, need to be written as part of the scenario materials available to EXCON. Scenario developers might think of the task as developing two sets of materials: one set is the standard background, context, and mission planning information required for any scenario-based exercise or wargame, and a second set provides additional scenario details, including not only important facts that aren't in the materials provided to the players, but also the story of how the operation will unfold and all of the details of what will happen to maneuver forces later in the game. See Section 3.4.5 for further details.

3.4.1. Required Level of Detail in Full Formation Plans for BLUE and RED

The two teams represent just the OIE working group (or equivalent) in the staff. All remaining staff sections are controlled by EXCON and represented to the players as abstractions or in the persons of the S2 and S3 role-players. This means that part of scenario preparation and maintenance includes the creation of all aspects of the RED and BLUE plans that are not related to OIE. That includes the base plan that OIE will support and the evolution of that plan as it unfolds during the game.

Maneuver plans from other staff sections built into the scenario need to be sufficiently detailed to allow the teams to plan OIE to support them, and maneuver plans need to be sufficiently detailed for the narrator to describe what happens in the operation at the end of each turn. An updated maneuver plan will need to be provided to each team at the beginning of each turn. These should be part of the pre-prepared materials that are part of the scenario.

Playtesting revealed the importance of having fairly elaborate and detailed plans from maneuver elements. When maneuver objectives and intended actions were vague, playtest players struggled to connect their proposed actions to aspects of the mission they were supposed to support. Because connection to operational objectives is one of the scoring criteria for evaluation and adjudication of actions, clear and detailed objectives and maneuver plans are required in order to be fair to the players.

3.4.2. Scenario and Plan Information Required for Adequate Preparation of the Role-Players and Adjudicators

The game calls for two sets of role-players: S2s and S3s for each side. To allow these role-players to function effectively, they need to be familiar with the details of the context, the details of the operation's progress to date (prior to each turn), and the intended progress of their side for the turn. Basically, **the role-players need to know everything the players need to know, but they also need to know more**.

Specifically, the role-players need to know the answers to the kinds of additional questions that the players will have so that the role-players can preserve credibility in their role by knowing more than the players. Some of this can be achieved by the role-players reading the scenario introductory material more carefully than the participants, but some of the information in the scenario should initially be EXCON-only material—things that players might (or should) ask as RFIs. Such material might include

current specific locations of friendly or enemy formations (especially IRCs), information about the personalities of formation leaders (on both sides), information about the status or intentions of GREEN groups, expected weather, etc.

Examples of material that might be useful for EXCON include the data tracked in Figures 8 and 9. Figure 8 shows an influence tracker developed by MCIOC for IWX 20.2. It allows for quick and holistic tracking of the level of influence each side has with or over various key groups, forces, or audiences within the scenario environment. Figure 9 depicts the summary assessment of the will to fight of the two scenario forces at the outset of the scenario broken down into component elements.[6] Tools like these could be used to help EXCON track important changes in the game.

Game aids such as the influence scoresheet or a will-to-fight tracker could also be displayed to the players to give them a better idea about where actions might be successful and the evolving influence landscape they face. If such summary tools are displayed to players, we recommend several caveats. First, the influence tracker should be modified to include an "audience TBD" column, to show that there may be TAs other than those listed on the scoresheet against which the teams could plan actions. Second, it should be made clear to the players that these scoresheets do not represent any intelligence or staff product that actually exists, and that during a real operation they would not have access to a definitive assessment of these factors and would likely have to assemble their own best guesstimate. Finally, such tools should be displayed only in the engagement room and should be noted to be (at least partially) covered by the "what happens in the engagement room stays in the engagement room" rule described in Section 2.5.4.

EXCON also needs to have fairly comprehensive tables of organization and equipment for both sides, not just for the IRCs available to each, but for the other capabilities, especially various different systems

---

6 For more on will to fight, see Ben Connable, Michael J. McNerney, William Marcellino, Aaron Frank, Henry Hargrove, Marek N. Posard, S. Rebecca Zimmerman, Natasha Lander, Jasen J. Castillo, and James Sladden, *Will to Fight: Analyzing, Modeling, and Simulating the Will to Fight of Military Units*, Santa Monica, Calif.: RAND Corporation, RR-2341-A, 2018.

FIGURE 8

Scoresheet Used to Track Team Influence with Different Audiences During IWX 20.2

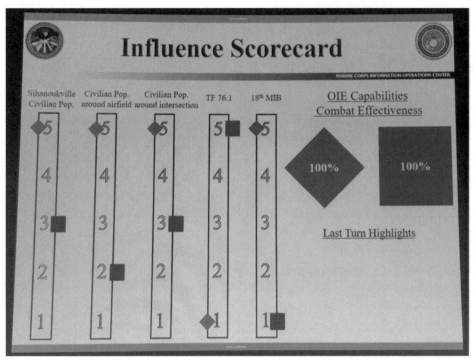

Photo credit: Nate Rosenblatt, RAND.

FIGURE 9

Will-to-Fight Summary Comparison of the Initial Status of the Two Opposed Forces in IWX 20.2

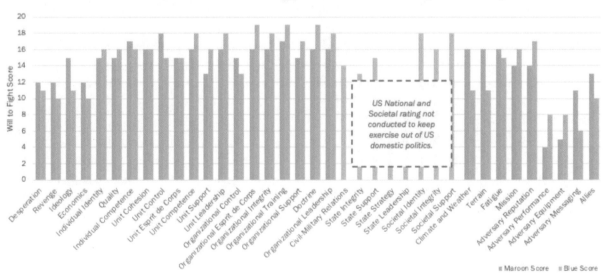

and platforms. Because the inherent informational aspects of all military activities can be leveraged as part of OIE, what capabilities are available is an important part of the scenario baseline. In addition, **EXCON needs detailed information about the performance and actual capability of available capabilities**.

During playtesting, creative players "made up" the operational effects possible from capabilities they were unfamiliar with; EXCON needs to be prepared to impose the real-world constraints that are inherent in available capabilities. (This could be the responsibility of an individual charged with the optional EXCON role of "reality master"; see Section 3.3). Ideally, sufficient expertise or pre-study of the capability by EXCON (perhaps by the narration team, the "reality master," or the EXCON element responsible for scenario design) will enable EXCON to provide needed answers when questions of capability come up.

If the answers aren't known, it creates a learning opportunity for both players and EXCON. Also, see Section 2.6—the show must go on. If getting to adequate information on the real-world capabilities of an IRC or other capability is going to take too long and disrupt the flow of the game, EXCON should impose an assumption (either their best expert judgment or something based on a die roll) so that the game can continue. (EXCON should also do their level best to track down the correct answer and share it with all participants as part of the game's after-action review to facilitate learning.)

3.4.3. Scenario Details Related to the Game Boards/Maps

This game is intended to be played with an associated scenario-specific set of "game boards" (which might be actual boards or might just be large maps printed with a plotter). The content of these maps needs to be provided as part of the scenario. The scale and size of the map should be determined based on the sizes of the two maneuver formations and the expected scope of the operation (both in terms of distance and of time). This requires some thought to make sure that the operation uses/unfolds across much of the board/map (it would be a shame to have the whole opera-

tions unfold within 4 square inches in the middle of a table-sized map) without having too much or any of the relevant action taking place "off map." (It is okay if some capabilities fly in from off map, or if naval fires are directed from off map, but ideally all of the action and the effects can be portrayed on the game board.)

In addition to the actual map or board, the scenario needs to include some details about the physical space represented by the map: the length of a key route, for example, and an estimate of how long it will take forces to transit it, or an estimate of throughput of local traffic. Information about the physical characteristics of the information environment could also be provided: the broadcast footprint of a TV station, the locations of cell towers, etc. Such information might be depicted as a "layer" that could not actually be laid on the game board, but that one of the S2s could use to pinpoint (and perhaps note with a token or marker) some key feature of the physical information environment.

The actual representation of the game board and markers or tokens is discussed further in Section 3.8.

3.4.4. Level of Detail Regarding GREEN Required in the Scenario

As has been mentioned, the scenario often needs to include details about culture, context, and local groups, often commonly referred to as GREEN (to distinguish them from the primary opponents, BLUE and RED). Because this is primarily an influence game, one or more groups under the purview of GREEN is likely to become a TA for actions by one or both teams. To support this, the players will need information about such groups, so such information should either be part of their intelligence briefing or something they can request as an RFI.

Details should include the type or composition of different groups or population segments, baseline behavioral proclivities, relevant cultural characteristics, significant narratives, views on RED and BLUE, relationships with other groups, leaders/influencers, media use patterns, etc. The more specific the information provided in the scenario, the better the teams will be able to tailor precise and effective actions (and presentation of those actions) to match, and the better the judges will be able to assess the prospects for success of such actions.

3.4.5. Scripting a Range of Operational Outcomes Possible at the Conclusion of Each Turn

As noted in Section 3.4, the scenario should include the story of how the whole operation will unfold except for the OIE parts. This is referred to elsewhere in the ruleset as the *storyline*. The initial scenario materials available to EXCON should have a likely storyline that covers the expected flow of the operation across all six turns of the wargame. The storyline needs to be somewhat flexible in case the teams' actions have a significant impact on progress toward one or more of the operational objectives. Even leaving room for that flexibility, the scenario should provide pre-packaged materials for the Step 1 update briefing for every turn that can be slightly adjusted or tailored based on any dramatic OIE effects from the previous turn or turns.

As noted in Section 2.6.1, the wargame has a range of possible outcomes for each turn, driven primarily by preplanned maneuver outcomes but capable of being affected by OIE along three tracks or storylines: (1) an expected direction of progress, (2) progress based on RED OIE outperforming, or (3) progress based on BLUE OIE outperforming. If teams and dice are relatively balanced, turn-by-turn progress of the overall operation should remain within these left and right bounds.

Depending on the time available for scenario development, the scenario should include a fully developed storyline and set of update briefings for track 1, the expected unfolding of the operation. Tracks 2 and 3 may be more or less developed. Ideally, all three storylines would be fully developed and the S2, narrator, or head judge could decide between Step 5 and the following Step 1 which of the three storylines the game is closest to and then quickly customize or tailor from there based on actions and their outcomes.

The overall level of detail in the storyline should correspond exactly to the level of detail needed to successfully complete Step 1 for each turn after turn 1. That is, there should be a prepared briefing covering all Step 1 topics waiting to be very slightly tailored based on the previous turn's Step 5 and presented to the teams so they can plan their next actions.

To develop the left and right bounding storylines, scenario designers should imagine possible effects that OIE might have on maneuver. It is difficult to imagine exactly what the teams will prepare, and impossible to anticipate which actions will be accompanied by sufficient dice luck to succeed. However, the range of possible effects and their impact on maneuver is easier to imagine. Many different information environment effects might keep an enemy formation from moving. That could be because of interrupted or falsified command networks, diminished will to fight, suppression, causing noncombatants to block routes, causing enemy forces to fear movement, causing enemy commanders to believe it is in their interest to remain in position— there are many possible ways to have the same effect on the game, namely, that a formation doesn't move or doesn't move as far as intended.

Opposite to actions that have the effect of slowing the progress of a formation (or a whole side) are actions that might speed the movement of a formation toward its goal. Actions that end up removing obstacles (getting a formation to move to a different location or retreat) or actions that reduce resistance (a feint that causes a formation to orient in the wrong direction, making it vulnerable to a flank attack, or successful efforts to keep refugees from clogging roads) could similarly speed progress. The basic outlines of left and right bounds for storylines, then, should probably focus on one condition where the progress of the attacking force (if the scenario contains such) is slowed or delayed, and another condition where the progress of the attacking force is easier than anticipated and accelerated. General outcomes along these lines could then be tailored to be described as a consequence of whatever the prominent effects were from the game actions of the previous turn.

NOTE: Many of the effects of OIE need not actually change the storyline: They can be *described* as having had the intended effects on the TAs and having enabled operational objectives without requiring the storyline progress to change. This is **not** to suggest that OIE do not have important effects that impact operations and campaigns! Quite the contrary, the authors are strong believers in the importance of effects in and through the information environment. However, just because OIE have effects doesn't mean they *actually* have to affect the flow of what happens across turns in the game, only that they be *described* as affecting outcomes and progress. The narrator or EXCON lead can keep the game exciting by describing the effects of an OIE action on the game, but the extent to which that action actually affects the game depends on their best judgment and the overall flow of the storyline.

Between this opportunity to narrate effects without actually changing pre-programmed progress on the storyline and left- and right-bounded alternative storylines, it should be possible to keep the game within these anticipated tolerances. Should one team's actions be so consequential as to start to push the storylines out of bound, perhaps it is time for a corrective inject. See Sections 2.6.1 and 2.6.1.2.

3.4.6. Preparing Scenario Injects

Section 2.6.1.2 provides a discussion of using injects in the game for three purposes: to stimulate players to more interesting actions, to challenge teams with unanticipated problems, and to increase difficulty for one team to correct the overall progress of the story-line and get it back within tolerances. Section 2.6.1.2 suggests that injects be presented as cards to both teams to create a dynamic feel within the game and give players something concrete to respond to.

Injects are not strictly necessary. Injects were limited in playtesting, and the games were still sufficiently dynamic with just the chaos caused by the two teams and their actions. Injects are optional.

Although injects should *feel* dynamic, they need not be. Injects can be a planned and preprogrammed part of the storyline but just presented as if they were dynamic, or they can actually be changes in what was originally planned. If the storyline includes some preplanned setback or misfortune for the maneuver progress of one side, we recommend that that be presented as if it were actually a dynamic inject!

Remember when designing injects that the goal is to stimulate or challenge the players, *not* to make things more difficult for EXCON. Injects that are likely to cause big swings in the storyline may seem like a good idea but are likely to be unnecessary and will make the game harder to run.

If injects are going to be available to be used, a set of injects should be prepared as possible inclusions with the storyline update materials for Step 1 of each turn. They should be cards that can be given to each team. Because each side may perceive the events of an inject differently, cards may need to be team/side specific. For example, a RED unit leader being executed when his infidelity with his commander's wife is exposed might have different indicators for BLUE or for RED, and, when it is reported in the media, the two sides may assign different levels of credibility to the report.

Injects need to either include or be backed up by sufficient detail. Think about the kinds of RFIs that an inject is likely to generate, what the actual answers are, and what each side's intelligence capabilities will be able to discern. Preparing this via playtesting is a good way to ensure you have that information ready at the start of the game.

For injects to be interesting, they need to either create an opportunity or a problem within the information environment. Also, injects should be written and described as if they are chance events. Injects should *not* be OIE from elsewhere in the force (a higher echelon, an adjacent formation). Injects should appear spontaneous and accidental. Some possible examples of injects:

- A key influencer is accidentally killed by the operations of one of the two sides.
- A key leader of part of one of the subordinate formations is killed in combat, or is relieved with cause, and either leaves that formation rudderless (and perhaps more vulnerable) or is replaced by a new leader who has different proclivities.

- International outrage descends on one side because of reports of atrocities (whether or not these accusations are true and who made them might be pointed RFIs).
- Something about the operation or the response leads to protest in area (peaceful or otherwise).
- A BLUE helicopter goes down, and recovery operations (and recovery support operations) become necessary.
- The presence of noncombatants in the area is much higher than intelligence suggested because of a new local shrine (or an unanticipated local holiday or festival, or some other reason), and this increased presence is a threat to operational progress for one or both sides.
- A BLUE vehicle hits and kills a small child; it is unclear what the response throughout GREEN will be.
- A commander or leader on one side is embroiled in some kind of personal scandal that breaks in the news today.
- Rolling power outages restrict internet and mobile phone availability across part of the area of responsibility (even where batteries are charged, the towers are dark).

3.5. Preparing Academic Instruction and Preliminary Planning

Playtesting has revealed that an academic period (for cross-leveling and instruction) and a planning period prior to the start of the wargame are essential to success.

At minimum, the academic period must include instruction regarding the terms that will be used and cross-leveling to baseline for participants to be able to meet expectations regarding plans for actions to be proposed.

Similarly, the time allocated for Step 2 in the flow of each turn of the wargame is not sufficient for participants to digest the scenario and plan an OIE concept of support. Such planning (and related familiarity with the scenario and maneuver plan) must be begun before the wargame begins. The better the quality of the teams' existing plans prior to the wargame, the better (and smoother) the game will go.

Total available planning time during the academic period, during all Step 2s and between turns, should be a consideration in determining how many actions to expect and allow from each team each turn. Experience during playtesting was that, with 60 minutes to prepare, teams will design two actions if they do not have existing plans prior to the wargame, and three actions if they do. If planning time is limited, fewer actions should be requested, allowed, and expected.

> **NOTE:** Having fewer proposed actions makes each team's prospects for any form of success in a given turn more susceptible to the roll of the dice.

3.6. Estimates of Timing for Play

As noted, turns are estimated to take between 3.5 and 6 hours to fully resolve. The width of variation in this estimate is due in part to lack of experience with this brand new wargame, but also to possible variation in time to be allocated for various player activities and variation in the number of actions per team per turn (see Section 3.7). Allowing too little time may leave players unprepared, whereas allowing too much time could make the game feel slow and unengaging. Table 8 summarizes activities by step and gives a recommended time allowed, as well as estimates of time that might be taken under slow or less than ideal circumstances. These time estimates are to support worst-case planning for game flow, or, should a compression of timeline be desired to fit more turns in a day, to provide estimates on steps that might be amenable to reduction in allocated time.

If there is a desire to increase turn speed (to either increase the total number of turns or reduce the number of hours spent in game play), Step 1 (and possibly Step 2) may be omitted on the first turn of the game, as Step 1 will have been fully covered in the first week of IWX, and Step 2 should come fairly directly from each team's prepared plans. Similarly, time might be shaved at the end or beginning of each day by allowing Step 5 to be completed after the conclusion of the exercise for the day, or by instructing teams to arrive the next day ready to immediately begin with Step 3 for the next turn.

TABLE 8

Estimates of Time Required Per Step, in Minutes

Step	Activity	Planned Time Limit	Lower Time Estimate	Maximum Time Estimate	Estimate Range
1	Receive situation and update	15	15	30	0–30
2	Prepare to present	60 or less	30	60	30–60
3	Present actions	45	3	60	60–90
3a	Final revisions to actions	15	15	30	
3b	EXCON prep for Step 4	Simultaneous			
4	Engagement meeting	90	90	120	90–120
5	Results and reset	45	15	45	15–45
Total		270			210–345

3.7. Choosing Target Number of Turns

Choosing the target number of turns to take place within the game is a function of the number of training days available and the scope of the operation or scenario intended to be covered by the wargame. The format for IWX allows three training days for the conduct of the wargame. The default assumed pace of play for the game is two turns per day, so in a three-day game, the natural default number of turns is six.

A number of considerations could lead to something other than this default of three days/six turns. With long days and a certain degree of compression of (and ruthlessness with) time allowed per step, it might be possible to complete three turns each day. This could allow a six-turn game to be completed in two training days. Or time for preparation before the game or for reflection after the game may be lacking. IWX includes preparatory briefings and the possibility of a demonstration turn prior to the game beginning and outside of the three training days ideally allocated to the game, and allows for an after-action review in the morning on the training day following the game.

If the total number of training days available to any wargame activities is tightly constrained, some of these activities might still be incorporated but at the cost of turns. Perhaps the first turn is a demonstration turn (which might still "count" in the game, or might just be a demonstration); similarly, perhaps the game could include only five turns and the afternoon slot on the third training day could be allocated to after-action review and clean-up.

Further considerations on number of turns may stem from the scenario. An abbreviated scenario that doesn't escalate to combat might instead require fewer turns (perhaps two shaping turns and two operational turns). Or a scenario might be more involved or cover a longer total time span with training objectives that require consideration of that more extensive scope. If that is the case, perhaps compressed turns (three per day) and/or more training days allocated to the wargame would be appropriate.

3.8. Preparing the Game Board and Various Representations

This game can be run in a single physical space or separated over virtual space.[7] In both cases, at least three isolated rooms will be required:

- BLUE side planning room
- RED side planning room
- central engagement room.

----

[7] Restrictions imposed by the 2020 COVID-19 pandemic required virtual/remote components in the design considerations from the ground up for the IWX wargame design. This game could be played virtually, or in a combined virtual-physical design that relies on video projection capability, the relaying of game maps to virtual participants, and the possibility of employing virtual gaming services. Any effort to shift some or all aspects of this game to a virtual platform would require considerable design and rehearsal on the front end, as well as redundant communications to ensure that no single point of communications failure interrupts the game flow.

These three rooms can be collocated in a single building or workspace or separated across a video teleconferencing system. This section describes both approaches. Whichever option is selected, the game also requires a fixed set of gaming materials, described below.

Game boards/maps: This is a tabletop game played on a flat map surface laid out on either a large table (preferable) or a floor. As designed, the game board consists of a map with military grid reference system (MGRS) markings at a scale appropriate to the scenario. *All three maps must be identical.* One will be provided to each of the two opposing teams, and one will be placed in the central engagement room. For the October 2020 IWX version of the wargame, MCIOC EXCON employed one map that was 172″ × 148″ at 1:25,000 tactical scale in the engagement room (shown in Figure 10), and a smaller map in each of the planning rooms (Figure 11). As of late 2020, MCIOC retains copies of these maps and the map data.

The specific sizes and scales of the maps can be adjusted to fit any selected scenario as long as they are identical to each other and have sufficient space to allow for planning, maneuver, and free play. Figure 11 shows the map used for the IWX 20.2 scenario.

Electronic map alternative: If the game is played entirely in the virtual realm, or replicated virtually, there are several options available for electronic game board depiction. None of these were tested during the IWX 2020 wargame playtest and development period, but they could be explored and tested for future iterations. Options include but are not limited to the use of the Office of the Secretary of Defense Standard Wargame Integration Facilitation Toolkit (SWIFT); the open source VASSAL virtual gaming system; or Google Maps, Google Earth Pro, or a similar shared mapping service. SWIFT has the capability to host classified gaming.

Physical tokens and icons: Units, population elements, and assets are represented by physical tokens, which are distinguished by their color and the icons on them. The IWX 2020 wargame used a common set of 2-inch-high plastic tokens set in 1-inch round bases. Tokens and icons are used to help players and EXCON visualize the position of units and assets on the map, and also to help track the availability of assets not yet in play. Red tokens (alternatively called *game pieces* or *markers*) represent the RED side units and have icons in diamond patterns, in line with standard operational terms and symbology. Blue tokens represent BLUE units and have icons in blue rectangles, also in line with standard practice. Black

FIGURE 10

Map Displayed in the Engagement Room During IWX 20.2

Photo credit: Nate Rosenblatt, RAND.

FIGURE 11
Map Displayed in the Planning Rooms During IWX 20.2

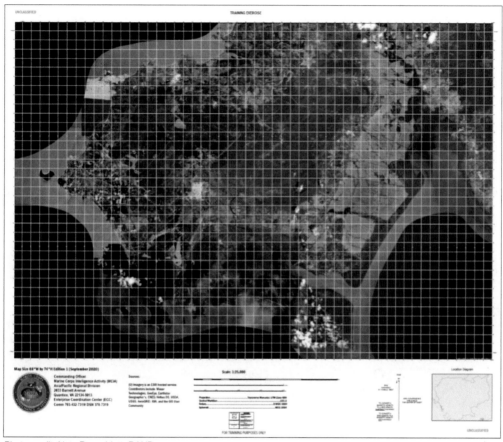

Photo credit: Nate Rosenblatt, RAND.

icons with white background symbolize civilian groups and different types of information activities. Figure 12 depicts the general icon design, with red, black, and blue icons presented from left to right. Figure 13 shows the icons used for BLUE and RED units, formations, and capabilities, and Figure 14 shows the icons used for noncombatants and infrastructure. Finally, Figure 15 shows some of the actual tokens as used in the IWX 20.2 wargame.

More information on operational symbols and graphics can be found in U.S. Army Doctrine Publication 1-02, *Terms and Military Symbols.*

Each specific scenario will require a different mix of icons. However, the set acquired by MCIOC for the IWX 2020 wargame would be suitable for adaptation to many different scenarios involving Marine Corps forces and information activities. The RAND team acquired this set of icons from Litko Game Accessories (www.litko.net). Costs have

FIGURE 12
General Icon Types Used in IWX 20.2

RED   Noncombatants   BLUE
and infrastructure

already been paid for up-front artwork, so subsequent orders should have reduced costs. Anyone wishing to reorder a set of the IWX 2020 icons from Litko should reference order number 919268. Several other companies produce similar custom wargaming icons.

Electronic icons: Each of these icons can be replicated in electronic format. Both SWIFT and

FIGURE 13

Icons Used for RED and BLUE Units, Formations, and Capabilities in IWX 20.2

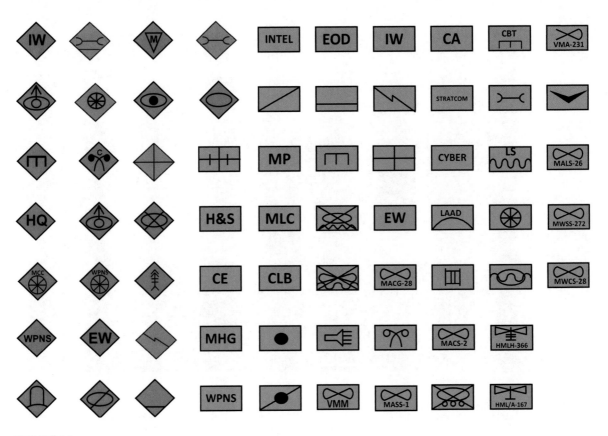

FIGURE 14

Icons Used for Noncombatants and Infrastructure in IWX 20.2

FIGURE 15
Physical Icons from IWX 20.2

Photo credit: Lance Corporal Kaleb Martin.

VASSAL support the replication of icons in virtual form.

Additional materials: Playtesting and experience with IWX 20.2 revealed that additional markers and representations could be valuable. These might include translucent plastic radius templates (that could be used to show broadcast radii, jamming radii, drop zones for leaflets, etc.), as well as large plastic arrows to denote intended lines of advance (so tokens could show current positions of maneuver formations and relevant GREEN organizations or groups with large arrows showing their intended directions of progress during the turn, adding more dynamism to the battlespace). These arrows could correspond to the major colors in the game, so blue, red, and green.

Also mentioned was the possible addition of "status rings"—plastic rings that can be added to the base of a token to denote some status of the unit (suppressed/pinned, out of communication, fleeing, resolved, etc.). These status rings might also be produced in a range of colors, and ring color might be used to denote the degree of influence one side or the other has over a unit (or civilian population group)—where no ring would denote normal/baseline degree of influence, a light red ring might denote some RED influence, and a bright red ring might denote more extensive RED influence (and the converse for BLUE), for example. Also, an observer suggested that flags or flag tokens could be made available and be used to denote BLUE or RED control over certain areas or key locations or facilities.

In addition, players in the wargame at IWX 20.2 found that having a pointer (a long stick, as depicted in Figure 5 on p. 31) was useful when briefing actions on the large game board/map.

4. OTHER CONSIDERATIONS RELATED TO THE CONDUCT OF THE IWX WARGAME

4.1. Consider Conducting a Preview or Demonstration Turn

To maximize player comfort with the steps and procedures described in this ruleset, it may be appealing to run a quick demonstration turn. A demonstration turn would likely be confined to Step 4 (Engagement) and perhaps a limited Step 5 (Results and Reset), assuming that Steps 1–3 have already been completed.

Ideally, the demonstration turn would be for a different scenario than the ones that the players will be playing so as to avoid giving them preconceived notions about appropriate actions in their scenario. If that is not possible (perhaps this is the first iteration of this wargame, or perhaps materials from previous scenarios are unavailable), then a sample turn from the current scenario will have to do.

The goal of the demonstration turn is to demonstrate the procedures. It is important that the demonstration illustrate the details of Step 4 as closely as possible to how they will actually be conducted during the game. This means that actions should be presented by demonstration players adhering to the rule requiring a different presenting player for each action, and that, after the matrix discussion, that player gets to make the outcome roll.

The EXCON lead (or the narrator) should act as a narrator for the demonstration turn, describing what is happening and why as the demonstration turn unfolds.

Materials for the demonstration turn should be prepared in advance and at least partially scripted so that demonstrators are not making things up on the fly. This scripting should include the details of the actions, the three reasons for action success, and the three rebuttal reasons offered by the opposing team (though the demonstrators for the opposing team should make a show of conferring and discussing before offering their scripted counterarguments).

These scripted pros and cons should template the types of arguments that are desired from players during the actual game—for example, arguments

that an action will fail because of elements in the scenario context (the TA is firm in their resolve and will not be so easily swayed) and arguments that an action will fail because of failures in execution (the acting team defined the TA poorly, and the intended message will be only weakly influential across three groups rather than strongly compelling to one).

Whether or not there is a demonstration turn, when the adjudication process is first introduced and explained to the players, take care to emphasize the process by which target numbers are developed: that there are scoresheets that capture the judges' expert judgment, and that the judges score the difficulty of the action, the demonstrated quality of planning, and the impact of the debate between the two teams as part of the process that determines the target number.

Some playtesters noted that the determination of target numbers felt like a black box with little relation to the matrix discussion; this was particularly the case when actions from both teams received similar target numbers but one of the actions had clearly been "better"—failing to recognize that one action was better planned and presented but was also more difficult, while the other action wasn't as well planned but was also easier. If players understand that the target numbers combine three different things—difficulty, planning, and debate performance—they are more likely to accept the target numbers as fair.

4.2. Network Drops and Management

Ideally, each of the spaces used for the game (RED and BLUE workrooms, the engagement room, and whatever space EXCON resides in, which could also be the engagement room) should have computer connectivity and projection capability. Ideally, this would be supported by shared drives so that players could prepare material in their workroom and then easily display it from a computer in the engagement room. If such shared drives are available, at least someone on EXCON should have permissions to all shared drives/folders both so that they can support display of each team's materials and so that EXCON can "spy" on the work-in-progress actions that the teams are developing. Such spying can help with pre-scoring of scoresheets and can help the narrator imagine and prepare different possible deviations of the storyline.

If drives and folders are shared, teams may need to be explicitly instructed *not* to look at materials belonging to the other team; although we encourage "spying" by EXCON, one team viewing another team's plans or proposed actions prior to their presentation during Step 4 would not be fair.

4.3. Probability Distribution for Rolling Three Six-Sided Dice

Table 9 shows the probability distribution for rolling and summing three six-sided dice. While initially used during playtesting to help calibrate the scoresheets and the target number formulae, playtest judges found the table to be a useful reference for understanding how likely teams were to succeed at actions with various target numbers.

TABLE 9

Probability Distribution for Rolling Three Six-Sided Dice

Target Number	% Chance of Reaching or Beating the Target Number	Target Number	% Chance of Reaching or Beating the Target Number
3	100	11	50
4	99.5	12	37.5
5	98	13	26
6	95	14	16
7	91	15	9
8	84	16	5
9	74	17	2
10	62.5	18	0.5

ANNEX. METHODS USED IN DEVELOPING THE IWX WARGAME

This annex describes the research and other efforts that went into the development and playtesting of this wargame. In a traditional RAND report, the documentation of methods and background would be placed more prominently. However, since the primary purpose of this report is to document the rules themselves, we relegate this (still important!) collection of details to an appendix.

The sponsor for this research, the Marine Corps Information Operation Center (MCIOC), sought RAND's help in developing a wargame for use as part of MCIOC's twice-annual Information Warfighter Exercise (IWX, formerly Combined Unit Exercise, or CUX). Traditionally, the two-week IWX/CUX has included a period of "opposed free play" during the second week, in which various groups of participants would expose their plans to a "murder board" of senior peers for critique or face ad hoc adjudication of the effectiveness of their plans versus the effectiveness of a RED team's plans during different phases of a notional scenario operation. Two of the authors had experience with these previous efforts at gamification: Jim McNeive, in his role at MCIOC, and Christopher Paul, during previous RAND work in support of MCIOC (including serving as a senior mentor during several iterations of CUX). While CUX and IWX benefited from the inclusion of a practical planning component with some stimulation of and friction for that plan, these efforts left a significant margin for improvement available as an exercise wargaming experience.

This, then, was the foundation of the requirement MCIOC sought to meet with RAND's support: a better wargame structure for the IWX that had formal elements of wargames, including clearly defined teams, time-constrained turns, defined actions, a structured adjudication process, and a clear set of rules to govern those processes. Coauthor and principal MCIOC point of contact Jim McNeive brought an initial vision of the game: a three-day/six-turn wargame focused on the influence mission of Marine Corps OIE in which two teams would each take a number of actions per turn that would be formally adjudicated using some form of matrix adjudication (described in greater detail below). Working with McNeive and with the support of many MCIOC personnel, the RAND team refined this vision into a codified ruleset, supported the playtesting of the ruleset over several iterations, and ultimately produced the ruleset included here. This annex documents that development process.

A.1. Background: Operations in the Information Environment

Operations in the information environment (OIE) is a relatively new term of art that emerged first within the 2016 Department of Defense (DoD) *Strategy for Operations in the Information Environment* and was echoed in the 2018 *Joint Concept for Operating in the Information Environment.*[8] The Marine Corps defines OIE as "actions taken to generate, preserve, or apply military information power in order to increase and protect competitive advantage or combat power potential within all domains of the operational environment."[9] The Marine Corps has placed new emphasis on OIE, designating information as a warfighting function, establishing a Deputy Commandant for information, and creating force structure for information within the Marine Expeditionary Forces (MEFs), the MEF information groups (MIGs).

Early writings and proto-doctrine establish Marine Corps OIE as having seven functions:[10]

- Assure enterprise command and control (C2) and critical systems.
- Provide IE battlespace awareness.
- Attack and exploit networks, systems, and information.
- Inform domestic and international audiences.
- Influence foreign TAs.
- Deceive foreign TAs.

---

[8] U.S. Department of Defense, *Department of Defense Strategy for Operations in the Information Environment*, Washington, D.C., June 2016; U.S. Joint Chiefs of Staff, *Joint Concept for Operating in the Information Environment (JCOIE)*, Washington, D.C., July 25, 2018.

[9] Eric Schaner, "What Are OIE?" *Marine Corps Gazette*, Quantico, Va., April 2020, p. 20.

[10] Schaner, p. 20.

- Control OIE capabilities, resources, and activities.

While all seven of these functions are amenable to treatment within a wargame,[11] the scope on intended training for IWX (and thus for the IWX wargame) emphasizes "influence foreign target audiences," with a secondary focus on "inform domestic and international audiences" and "deceive foreign target audiences." "Attack and exploit networks, systems, and information" might be included in the wargame, and current rules could cover adjudication of such functions; including the other functions might require some improvisation or adjustments to the rules.

A.1.1. Background: Information-Related Capabilities

One of the key insights of the *Joint Concept for Operating in the Information Environment* is that everything the joint force says or does has the potential to create or affect information, and so the joint force should deliberately leverage "the inherent informational aspects of military activities" as part of OIE.[12] The concept avoids legacy language regarding *information-related* capabilities, since the employment of all capabilities has inherent informational aspects. Regardless, there are certain capabilities that are explicitly and primarily intended to generate effects in and through the information environment. These are traditionally referred to as information-related capabilities (IRCs). We note "traditionally," as the latest revision of Joint Publication 3-0, *Joint Operations*, instead describes "joint force capabilities, operations, and activities for leveraging information."[13] Under that heading, JP 3-0 lists the following items: key leader engagement (KLE), public affairs (PA), civil-military operations (CMO), military deception (MILDEC), military information support operations (MISO), operations security (OPSEC), electronic warfare (EW), combat camera (COMCAM), space operations, special technical operations (STO), cyberspace operations, DoD information network operations, cyberspace-enabled activities, and commander's communication synchronization.

Marine Corps doctrine still uses IRCs but intentionally does not provide a list; the definition of the term is open ended, such that capabilities outside those that could be listed might be IRCs under certain circumstances (like the *Joint Concept for Operating in the Information Environment*'s "inherent informational aspects"). An IRC is "a tool, technique, or activity employed within a dimension of the information environment that can be used to create effects and operationally desirable conditions."[14] Although there is no doctrinal list, informal lists abound within the Marine Corps. One SME interviewed for a previous RAND project said they had seen charts listing as many as 30 distinct IRCs.[15] Another respondent for that project provided a slide that included the following list of "information environment activities" for the Marine Corps: intelligence, C2, cyberspace operations, MISO, CMO, space operations, electromagnetic spectrum operations, communications strategy and operations, OPSEC, information assurance, physical security, STO, KLE, defense support to public diplomacy, physical attack, and MILDEC.[16] That same individual added signature management (SIGMAN) as a notable omission to the list, probably because it is a nascent concept and capability.

Because the Marine Corps still describes these capabilities as IRCs, we use that term in this wargame. Because the Marine Corps intentionally declines to provide a definitive list of IRCs, this wargame does not do so either. Further, the adjudication mechanism has been designed to be able to accommodate the informational impact or other effects of any capability.

---

[11] See, for example, Christopher Paul, Yuna Huh Wong, and Elizabeth M. Bartels, *Opportunities for Including the Information Environment in U.S. Marine Corps Wargames*, Santa Monica, Calif.: RAND Corporation, RR-2997-USMC, 2020.

[12] U.S. Joint Chiefs of Staff, 2018, p. 1.

[13] Joint Publication 3-0, 2018, p. III-22.

[14] Joint Publication 3-13, *Information Operations*, Washington D.C.: U.S. Joint Chiefs of Staff, incorporating change 1, November 20, 2014, p. I-3.

[15] See p. 24 in Paul, Wong, and Bartels, 2020.

[16] Provided during an interview between a Marine Corps civilian and Christopher Paul, April 16, 2018; originally cited in Paul, Wong, and Bartels, 2020.

A.2. Background: Will to Fight

Will to fight is *the disposition and decision to fight, act, or persevere when needed*. Marine Corps doctrine centers on the idea that will to fight is the single most important factor in war: War is, ultimately, a contest of opposing, independent, and irreconcilable wills. Will to fight is also very hard to understand.

Ideally, Marine Corps OIE would be able to quickly and effectively assess enemy will to fight and use this assessment to pinpoint critical vulnerabilities for targeting and to help assess the impact of information actions. If the ultimate objective in war—and also in competition—is to get the adversary to stop fighting or competing, then a successful attack on their will to fight may be a battle- or war-winning attack. OIE are particularly useful in attacking will to fight.

RAND researchers developed an analytic model of will to fight that we applied in the IWX.[17] This model helps the user examine the factors we found to be most relevant to the will to fight of both adversary military units and friendly units. Defending will to fight also requires some self-examination to identify and shore up vulnerabilities.

There are 29 factors in the RAND Will-to-Fight Model and nine additional contextual factors. The 29 factors range from individual-level training and ideology to unit cohesion and leadership, organizational integrity and support, civil-military relations, and popular support at the national level. Table 10 lists the 29 factors and their associated subfactors.

Assessment of will to fight can be done through tools developed by RAND to support the U.S. military. For the 2020 IWX, the RAND team used the tools to conduct a will-to-fight assessment for both the BLUE and RED forces used in the scenario. These assessments identified critical vulnerabilities that could be exploited by players participating in the IWX.

Results from an effort to attack adversary will to fight should be compared against the will-to-fight assessment conducted prior to the IWX. For example, if the assessment shows (as this one did) a potential weakness in unit leadership and the relationship between unit leaders and marines in one unit, and the RED side attacks that relationship with an information activity, then they should be rewarded with a useful result. If a team fails to use the assessment and attacks into an adversary strength, then they should not be rewarded.

Will to fight can be attacked via information activities and kinetic activities, and ideally via both. The combination of psychological operations leaflet drops and bombing in the 1991 Gulf War contributed to the surrender of thousands of Iraqi Army soldiers.

Not all successful will-to-fight attacks will result in a unit breaking and running. Results are often more nuanced. For example, the unit may hesitate (which can be represented in a wargame as losing a turn of action). The unit may waver in its attacks, affecting its chances of kinetic success. In some cases, loss of will to fight may be temporary and can be recovered over time.

Tracking will to fight in a wargame can be done in several ways. One of the most useful is to have a will-to-fight tracker for each unit. RAND gaming teams found that a 1–20 tracker, with 1 being broken will and 20 being elite will, was most useful and provided the greatest variety of action and reaction.

A well-designed scenario should, ideally, center on the will to fight of each opposing team. A top-level organizational or national will-to-fight tracker can be used to determine overall victory conditions. For example, if two units are defeated and one unit breaks and runs, that side might lose three will-to-fight points out of 20, pushing it closer to defeat. A battlefield success or a successful inward-focused information activity might help recover one or more of those points.

For more information on will to fight, and on the use of will to fight for gaming, see Ben Connable et al., *Will to Fight: Analyzing, Modeling, and Simulating the Will to Fight of Military Units*, Santa Monica, Calif.: RAND Corporation, 2018. More information on will to fight can be found by searching for "will to fight" on the RAND website: www. rand.org.

---

[17] Ben Connable, Michael J. McNerney, William Marcellino, Aaron Frank, Henry Hargrove, Marek N. Posard, S. Rebecca Zimmerman, Natasha Lander, Jasen J. Castillo, and James Sladden, *Will to Fight: Analyzing, Modeling, and Simulating the Will to Fight of Military Units*, Santa Monica, Calif.: RAND Corporation, RR-2341-A, 2018.

TABLE 10

Factors in the RAND Will-to-Fight Model

Level	Category	Factors	Subfactors
Individual	Individual Motivations	Desperation	
		Revenge	
		Ideology	
		Economics	
		Individual Identity	Personal, Social, Unit, State, Organization, Society
	Individual Capabilities	Quality	Fitness, Resilience, Education, Adaptability, Social Skills, Psychological Traits
		Individual Competence	Skills, Relevance, Sufficiency, Sustainability
Unit	Unit Culture	Unit Cohesion	Social Vertical, Social Horizontal, Task
		Expectation	
		Unit Control	Coercion, Persuasion, Discipline
		Unit Esprit De Corps	
	Unit Capabilities	Unit Competence	Performance, Skills, Training
		Unit Support	Sufficiency, Timeliness
		Unit Leadership	Competence, Character
Organization	Organizational Culture	Organizational Control	Coercion, Persuasion, Discipline
		Organizational Esprit de Corps	
		Organizational Integrity	Corruption and Trust
	Organizational Capabilities	Organizational Training	Capabilities, Relevance, Sufficiency, Sustainment
		Organizational Support	Sufficiency, Timeliness
		Doctrine	Appropriateness, Effectiveness
		Organizational Leadership	Competence, Character
State	State Culture	Civil-Military Relations	Appropriateness, Functionality
		State Integrity	Corruption, Trust
	State Capabilities	State Support	Sufficiency, Timeliness
		State Strategy	Clarity, Effectiveness
		State Leadership	Competence, Character
Society	Societal Culture	Societal Identity	Ideology, Ethnicity, History
		Societal Integrity	Corruption, Trust
	Societal Capabilities	Societal Support	Consistency, Efficiency

SOURCE: Connable et al., 2018.

A.3. Background: Importance of Wargaming to the Marine Corps

Both DoD and the Marine Corps have renewed their interest in wargaming in recent years. The Marine Corps in particular is poised to invest a considerable amount of resources into improving service-level wargaming capability and increasing its number of annual wargames. Wargaming also remains an important tool for operators and for policymakers for learning, exploring, and thinking through potential consequences of planned or conceivable operations. Wargaming is an established tool in the defense and intelligence communities and is especially salient when planners are faced with difficult, complex problems and uncertain futures. For these reasons, resilient inclusion of the information environment and related considerations and effects in wargaming is important for the Marine Corps.

Indicative of a broader, revived interest in wargaming even before some of the more recent high-level focus on wargaming, in the past decade both the Army and Navy war colleges published wargaming handbooks that sought to better describe and instruct on wargaming practice.[18] Other signs of recent, renewed DoD interested in wargaming include high-level memos on wargaming by the Deputy Secretary of Defense and Secretary of the Navy,[19] a new Defense Wargaming Alignment Group (DWAG),[20] the creation of a DoD wargame incentive fund,[21] and DoD-sponsored wargaming conferences.[22] The Military Operations Research Society (MORS) created a wargaming certificate program in 2017 in response to the increased demand for wargamers within the defense community.[23] DoD interest in wargaming has also coincided with renewed interest in wargaming by other countries. The UK Ministry of Defence published its own wargaming handbook in 2017 that, unlike the Army and Navy wargaming handbooks, became doctrine.[24] China has also invested in computerized wargaming over the past decade, beginning with wargaming strategic problems but also expanding to interservice wargames and tactical simulations.[25]

Within this environment, the Marine Corps is expecting to increase its wargaming capabilities to better prepare for future combat. The Marine Corps not only expects to increase its volume of wargames, but also to increase the technological sophistication of the wargaming it currently conducts.[26] Marine Corps Systems Command is currently overseeing the development of a "world-class" wargaming capability that seeks to be data-enabled and analytically rigorous, incorporating computerized modeling and simulation (M&S) and using in-stride game adjudication.[27] Former Commandant of the Marine Corps General Robert Neller spoke about his desire for a "Star Trek–like holodeck" for wargaming.[28] Stakeholders within the Marine Corps wargaming community often express the desire for more sophisticated adjudication, visualization, analysis, and knowledge management for future Marine Corps wargaming.

Marine Corps wargaming constitutes a very broad set of activities. These include training events and simulations, discussion groups and seminars, planning exercises, reviews of plans, and course of action (COA) wargaming as part of the Marine Corps Planning Process (MCPP).[29] Other wargame approaches with a heavier emphasis on adjudication that are currently used by the Marine Corps include matrix games, hex-and-counter games, commercial computer games, commercial board games, and manual games used in combination with M&S and analysis. Wargames are used from the tactical level to the service level and above, with wargames dedicated to informing Marine Corps Title 10 responsibilities to organize, train, and equip the force.[30] Stakeholders involved in wargaming are engaged in activities such as concept development, capabilities development,

[18] James Markley, *Strategic Wargaming Series Handbook*, Carlisle, Pa.: U.S. Army War College, Center for Strategic Leadership and Development, July 2015; and Shawn Burns, *War Gamers' Handbook: A Guide for Professional War Gamers*, Newport, R.I.: U.S. Naval War College, undated.

[19] Bob Work, Deputy Secretary of Defense, "Wargaming and Innovation," memorandum for service principals, Washington, D.C., February 9, 2015; and Ray Maybus, Secretary of the Navy, "Wargaming," memorandum for Chief of Naval Operations and Commandant of the Marine Corps, Washington, D.C., May 5, 2015.

[20] Bob Work and Paul Selva, "Revitalizing Wargaming Is Necessary to Be Prepared for Future Wars," *War on the Rocks*, December 8, 2015.

[21] Garrett Heath and Oleg Svet, "Better Wargaming Is Helping the U.S. Military Navigate a Turbulent Era," *Defense One*, August 19, 2018.

[22] Phillip Pournelle, ed., *MORS Wargaming Special Meeting, October 2016, Final Report*, Alexandria, Va.: Military Operations Research Society, 2017, p. 5; Phillip Pournelle and Holly Deaton, eds., *MORS Wargaming III Special Meeting, 17–19 October 2017, Final Report, April 2018*, Alexandria, Va.: Military Operations Research Society, 2018, p. 2.

[23] Military Operations Research Society, "Certificate in Wargaming," undated.

[24] United Kingdom Ministry of Defence, Development, Concepts and Doctrine Centre, *Wargaming Handbook*, Beichester, UK: LCSLS Headquarters and Operations Section, August 2017.

[25] Dean Cheng, "The People's Liberation Army on Wargaming," *War on the Rocks*, February 17, 2015.

[26] Todd South, "Marine Wargaming Center Will Help Plan for Future Combat," *Marine Corps Times*, September 19, 2017.

[27] Program Manager Wargaming Capability, Marine Corps Systems Command, "PM Wargaming Capability," September 26, 2018, pp. 1–2.

[28] James Clark, "The U.S. Marine Commandant Wants a 'Star Trek'–Style Holodeck for Wargaming," *The National Interest*, September 30, 2017.

[29] MCWP 5-10, 2010, pp. 4-2 to 4-3.

[30] U.S. Marine Corps Warfighting Laboratory Future Directorate, "Title 10 Wargaming," undated.

training and education, science and technology development, operational planning, and others.

Independent of the widespread and institutionalized position that wargaming has in the present-day Marine Corps, when used properly, the method itself can assist in critical thinking, individual and organizational learning, and trial-and-error exploration of new concepts and warfighting approaches without costing lives or material. This is particularly true as the Marine Corps continues its transition from over a decade of counterinsurgency operations to other forms of warfare that are markedly different but for which the current generation of marines may have no firsthand knowledge. Wargaming thus has the potential to play an important role particularly in OIE, where concepts and understanding are nascent and in development but where the chances for experimentation in the real world may be very limited. Although wargames do not prove or "validate" concepts and approaches, they can be used to teach principles, offer perspectives on what does not work, and create additional insights into an issue. They have the potential to raise questions and potential consequences that have not previously occurred to participants.[31]

A.4. Background: Wargame Adjudication

One of the central elements in a wargame is the approach to or mechanism for adjudication. How will the outcomes of actions taken by the players be determined within the wargame scenario? Adjudication is the procedure to impartially resolve the outcome of interactions between sides in a game.[32] The wargaming literature acknowledges three basic kinds of adjudication:

- Free adjudication: The results of interactions are determined by the adjudicators in accordance with their professional judgment and experience.[33] The opposing sides reaching a consensus on the likely outcome of a nonkinetic interaction or engagement is a useful adjudication method in an open or mixed open/closed wargame format.
- Rigid adjudication: The results of interactions are determined according to predetermined rules, data, and procedures such as combat models or a combat results table.[34]
- Semi-free adjudication: a hybrid approach in which interactions are evaluated by something akin to the rigid method, but the outcomes can be modified or overruled by the lead adjudicator.[35]

The different types of adjudication methods have different strengths and weaknesses, and some are better or more poorly suited to different game purposes, types, and styles. For example, rigid adjudication is all but impossible where the types of actions sides might take are difficult to determine in advance or where the outcomes of such actions are contingent on numerous complex factors and their interactions. Such games are better suited to some form of free adjudication. However, free adjudication can be (or can appear to be) biased by the perspectives of the expert judges. For games involving a discrete number or types of actions or involving combat at a relatively high level of abstraction, a more rigid adjudication method is more appropriate. When determining the outcome between two clashing forces, a list of possible modifiers could be consulted, and then a probabilistic determination (through dice or some other random number generation system) could be made based on a combat resolution matrix developed based on historical combats of that type.

The IWX wargame uses a hybrid adjudication model that should be considered a form of semi-free adjudication. The process (as described in the rules regarding Step 4) begins with the presentation and description of the action, followed by a rebuttal from the other team, followed by a counterargument. This portion of the adjudication process is typical of what is called a *matrix game* or *matrix adjudication process*, which is a form of free adjudication.[36] However,

[31] Burns, undated, pp. 3–4.

[32] Burns, undated, p. 51.

[33] Francis J. McHugh, *U.S. Navy Fundamentals of War Gaming,* New York: Skyhorse Publishing, 2013.

[34] McHugh, 2013.

[35] McHugh, 2013.

[36] John Curry and Tim Price, *Matrix Games for Modern Wargaming: Developments in Professional and Educational Wargames,* History of Wargaming Project, 2014.

in this game, the matrix discussion informs a more structured and rigid step in the adjudication process: A panel of judges completes scoresheets, which lead to an assessment of the likelihood of success of the action (the target number), which is then subjected to a probabilistic determination (the outcome roll). Following this semi-rigidity of scoring and rolling dice, the adjudication turns back toward free with the description of that dice-determined outcome described by the narrator in a free format.

A.5. From Initial Vision to Playable Game

Within this context characterized by increased emphasis on the information environment, new concepts related to OIE, emerging research related to will to fight, and existing practice related to wargaming, the RAND team sought to transform the sponsor's initial vision into a playable wargame. Note that the authors brought considerable expertise to this effort. Some are hobby wargamers who have also been involved in the design and conduct of defense-related wargames (Connable and Paul). Some have considerable experience with information operations/OIE and related research (Paul and McNeive), and one recently served a tour as the information operations officer for Marine Forces Europe and Africa (Welch). Several authors' recent RAND research relates to the topic at hand.[37] These experiences informed refinement and specification from the initial concept developed by McNeive into an initial draft of the rules, which were then further refined through iterative discussions with MCIOC personnel. Once the rules and procedures reached a certain level of ripeness, the process of rehearsal and playtesting (described in the next section) began. While the initially developed core concept and structure survived all of the rigors of playtesting, playtesting did lead to continued evolution, minor adjustments, and additional optional rules, as well as refinements in the various supporting play aids. The rules reported here include improvements from all playtesting, including the actual execution of the wargame during IWX 20.2 as a final playtest.

A.6. Playtesting and Draft History

Starting with the initial draft ruleset developed by the RAND team building on the kernel of the idea from the sponsor, the game development process involved iterative testing and evolution of the rules through a series of rehearsals, practices, and playtests.

The 0.1 draft of the ruleset was completed on May 29, 2020. The 0.2, 0.3, and 0.4 drafts followed direct comment by MCIOC personnel on the initial draft and a series of phone discussions between the RAND team in early to mid-June 2020, with the 0.4 draft being subject to the first walkthrough playtest on June 23, 2020. The walkthrough was a half-day evolution with a partial EXCON and a single representative player who presented two pre-scripted actions just to give a feel for how the game would flow. This walkthrough proved to be a very useful experience in that it confirmed the feasibility of the general flow of the game as envisioned and exposed a wider range of the MCIOC personnel who would ultimately serve as EXCON members to the vision for the game. This evolution also led to the identification of numerous additional requirements for the rules, especially in the area of play aids: The June 23 walkthrough led to the first drafts of scoresheets for the judges (final versions now in the game materials available for download at www.rand.org/T/TLA495-1), the first draft "turn map" (final version now Figure 2), the first iteration of the "how to make an outcome roll" graphic (final version now Figure 1), and the first version of the IWX *Player's Guide* (the final version is available at www.rand.org/T/TLA495-1).

The next major playtest event took place on July 21 and 22, 2020. The goal of this evolution was to prepare and play two full notional turns of the wargame. A four-member BLUE team and three-member RED team (both recruited from within MCIOC) were asked to prepare several actions and were given draft scenario materials prepared by MCIOC personnel. The emphasis on this playtest (using ruleset 0.5) was on experiencing the flow and process of the adjudication step, Step 4. Key takeaways from this round of playtesting included the addition of a rebuttal to each discussion cycle (before

---

[37] See, for example, Paul, Wong, and Bartels, 2020; and Connable et al., 2018.

that point, the discussion was simply presentation and counterargument) and the identification of additional optional EXCON roles, such as an individual with responsibility for representing the GREEN perspective. Discussion subsequent to this playtest led to further adjustments and additions to the rules as well as the addition of an alternative approach to judge scoring (both are included in *GA8: Engagement Scoresheet*).

August 12, 2020, brought a scheduled "mini-rehearsal" using the 0.6 rules and focused on the timing and narration for Steps 4 and 5. This event helped solidify the target timer allocations to keep Step 4 running smoothly and to prepare the MCIOC EXCON team for the full playtest as part of IWX 20.1.

IWX 20.1 took place in August 2020 at Camp Lejune and involved members of the II MEF MIG as the BLUE side training audience; MCIOC SMEs played as RED, MCIOC provided EXCON, and the exercise and wargame used the II MEF MEFEX scenario. Some MCIOC personnel were live in person at Camp Lejune, but much of the support came via Secure Video Teleconference from MCIOC at Marine Corps Base Quantico due to COVID-19 restrictions. Because of constraints on connectivity, RAND personnel were unable to observe this playtest and so instead relied on notes and observations from MCIOC personnel. The 0.6 rules were used for IWX 20.1.

The RAND team did receive copies of all of the completed scoresheets used throughout IWX 20.1 and was able to use them to make adjustments to the scoresheets, including clarifying instructions, clarifying the phrasing of some of the criteria, and calibrating the target number generation process. These and other adjustments would be captured in the 0.7 ruleset used for the September 23, 2020, test run at MCIOC. This test run again included two turns over a single day. The focus was on making sure that the MCIOC personnel who would be part of EXCON for IWX 20.2 were prepared to do so and on cementing scoring procedures. This test run was considered to be highly successful, and the changes and suggestions that emerged were relatively minor, described as being of the "happy to glad" sort. The 0.7 ruleset was the first to included handouts with instructions for each of the major EXCON roles (the judges, the S3 role-player, and the S2 role-player—final versions included in the downloadable game materials).

Ruleset 0.8 was used in the first full IWX wargame, which took place over three days during the week of October 26, 2020. RAND personnel were able to observe in person and on the phone throughout the course of the wargame, and MCIOC collected after-action comments from participants and EXCON. These observations and input led to some final adjustments to the ruleset, including the development of a more substantial section of guidance for narration of events and outcomes to increase consistency in this important aspect of game adjudication. The RAND team captured this requirement and other observations and opportunities for improvement in a ten-page project report to MCIOC. Adjustments to the rules to take advantage of these remaining opportunities for improvement led to the 1.0 version of the rules as captured in this document.

A.7. Looking to the Future

This wargame is suitable for tactical- to operational-level wargaming of the OIE component of military operations across the spectrum from competition to conflict. With some modification, it could be used in a wide variety of contexts. We have written the rules with many noted optional rules or rules variants to allow EXCON to tailor the rules for specific events (IWX or otherwise) to best support their specific activity or exercise.

The IWX 20.2 wargame was a success. And, as a final playtest, it contributed further fine-grain improvements to the ruleset. MCIOC intends to use this ruleset for future IWXs and to modify the ruleset for use in an operational-level wargame to support an exercise for a Marine Corps geographic component command outside the context of IWX. Drawing on lessons learned from this and other applications of the game, we may publish supplements to this ruleset with considerations relevant to specific contexts, different staffing relationships, or different OIE problem sets.

REFERENCES

Burns, Shawn, *War Gamers' Handbook: A Guide for Professional War Gamers*, Newport, R.I.: U.S. Naval War College, undated.

Cheng, Dean, "The People's Liberation Army on Wargaming," *War on the Rocks*, February 17, 2015.

Clark, James, "The U.S. Marine Commandant Wants a 'Star Trek'–Style Holodeck for Wargaming," *The National Interest*, September 30, 2017.

Connable, Ben, Michael J. McNerney, William Marcellino, Aaron Frank, Henry Hargrove, Marek N. Posard, S. Rebecca Zimmerman, Natasha Lander, Jasen J. Castillo, and James Sladden, *Will to Fight: Analyzing, Modeling, and Simulating the Will to Fight of Military Units*, Santa Monica, Calif.: RAND Corporation, RR-2341-A, 2018. As of November 20, 2020: https://www.rand.org/pubs/research_reports/RR2341.html

Curry, John, and Tim Price, *Matrix Games for Modern Wargaming: Developments in Professional and Educational Wargames*, History of Wargaming Project, 2014.

Heath, Garrett, and Oleg Svet, "Better Wargaming Is Helping the U.S. Military Navigate a Turbulent Era," *Defense One*, August 19, 2018.

Joint Publication 3-0, *Joint Operations*, Washington, D.C.: U.S. Joint Chiefs of Staff, incorporating change 1, October 22, 2018.

Joint Publication 3-13, *Information Operations,* Washington D.C.: U.S. Joint Chiefs of Staff, incorporating change 1, November 20, 2014.

Marine Corps Warfighting Publication 5-10, *Marine Corps Planning Process*, U.S. Marine Corps, as amended May 2, 2016.

Markley, James, *Strategic Wargaming Series Handbook*, Carlisle, Pa.: U.S. Army War College, Center for Strategic Leadership and Development, July 2015.

Maybus, Ray, Secretary of the Navy, Department of Navy, U.S. Department of Defense, "Wargaming," memorandum for Chief of Naval Operations and Commandant of the Marine Corps, Washington, D.C., May 5, 2015.

McHugh, Francis J., *U.S. Navy Fundamentals of War Gaming*, New York: Skyhorse Publishing, 2013.

Military Operations Research Society, "Certificate in Wargaming," undated. As of May 13, 2021: https://www.mors.org/Events/Certificates/Certificate-in-Wargaming

Paul, Christopher, Yuna Huh Wong, and Elizabeth M. Bartels, *Opportunities for Including the Information Environment in U.S. Marine Corps Wargames*, Santa Monica, Calif.: RAND Corporation, RR-2997-USMC, 2020. As of December 3, 2020: https://www.rand.org/pubs/research_reports/RR2997.html

Pournelle, Phillip, ed., *MORS Wargaming Special Meeting, October 2016, Final Report*, Alexandria, Va.: Military Operations Research Society, 2017.

Pournelle, Phillip, and Holly Deaton, eds., *MORS Wargaming III Special Meeting, 17–19 October 2017, Final Report, April 2018*, Alexandria, Va.: Military Operations Research Society, 2018.

Program Manager Wargaming Capability, Marine Corps Systems Command, "PM Wargaming Capability," September 26, 2018.

Schaner, Eric, "What Are OIE?" *Marine Corps Gazette*, Quantico, Va., April 2020.

South, Todd, "Marine Wargaming Center Will Help Plan for Future Combat," *Marine Corps Times*, September 19, 2017.

U.S. Army Doctrine Publication 1-02, *Terms and Military Symbols*, Washington, D.C.: Headquarters. Department of the Army. Washington, D.C., August 14, 2018.

U.S. Department of Defense, *Department of Defense Strategy for Operations in the Information Environment*, Washington, D.C., June 2016.

U.S. Joint Chiefs of Staff, *Joint Concept for Operating in the Information Environment (JCOIE)*, Washington, D.C., July 25, 2018.

U.S. Marine Corps Warfighting Laboratory Future Directorate, "Title 10 Wargaming," undated.

United Kingdom Ministry of Defence, Development, Concepts and Doctrine Centre, *Wargaming Handbook*, Beichester, UK: LCSLS Headquarters and Operations Section, August 2017.

Work, Bob, and Paul Selva, "Revitalizing Wargaming Is Necessary to Be Prepared for Future Wars," *War on the Rocks*, December 8, 2015.

Work, Bob, "Wargaming and Innovation," memorandum from the Secretary of Defense for Service Principals, Washington, D.C., February 9, 2015.

ACKNOWLEDGMENTS

The development of the Information Warfighter
Exercise (IWX) wargame was a collaborative effort.
The contribution of the sponsor principal point of
contact, Jim McNeive, was so significant that he is
rightly included as a co-author. Special thanks to
the Marine Corps Information Operations Center's
(MCIOC's) Brian Doty, who had a leading role in the
planning and execution of the games and playtests.
Many other personnel at MCIOC made valuable
contributions as playtesters, in designing the scenario
and context in which the October 2020 version of
IWX (20.2) took place, and as judges and members of
exercise control during playtesting and actual execution; there are too many to acknowledge everyone
individually, but all are truly appreciated, nonethe-
less. We are also grateful to the personnel from
other organizations who took part in the playtests
or the actual IWX 20.2 game, including personnel
from the School of Advanced Warfighting at Marine
Corps University, staff from the office of the Deputy
Commandant for Information's office, personnel
from II Marine Expeditionary Force Information
Group, and the training audience for IWX 20.2 itself,
including personnel from across the joint force, as
well as a handful of international participants. We
further thank Maria Falvo for her administrative
support and for helping with formatting for this and
other project documentation. We are indebted to Bill
Marcellino and Andrew Lotz for their thoughtful
and helpful reviews. Fadia Afashe and James Torr
helped bring this report/ruleset into the final published form, no mean feat given that game rules differ
in style and substance from what is typical in RAND
reports. Rick Penn-Kraus created the excellent cover
design. Thank you all.

ABOUT THIS DOCUMENT

This document provides details and documentation related to the RAND project "MCIOC Information Warfighter Exercise Wargame Support." The goals of the project were to support the Marine Corps Information Operations Center (MCIOC) in the development of a wargame component for the calendar year 2020 Information Warfighter Exercises (IWXs) and to codify rules and guidance to include the same style of wargame in future IWXs or similar contexts.

With that latter goal in mind, this document presents the full ruleset for the IWX wargame, including a host of optional rules to allow tailoring the game to specific preferences, needs of the training audience, or scenarios. With the emphasis of the document on presenting guidance for the conduct of the game itself, details on the methods and process by which the game was developed have been relegated to an annex. Interested readers should see the annex on pp. 72–79 for context and methodological and development details.

Numerous handouts and aids for playing the game, as well a *Player's Guide*, are available at www.rand.org/t/TLA495-1.

This rulebook should be of interest to those who wish to conduct (or participate in the administration of) an IWX wargame at MCIOC. The rulebook should also be of interest to anyone wishing to conduct this wargame as part of a different training exercise series or anyone wishing to design a wargame related to operations in the information environment.

This research was sponsored by MCIOC and conducted within the Navy and Marine Forces Center of the RAND National Security Research Division (NSRD), which operates the National Defense Research Institute (NDRI), a federally funded research and development center sponsored by the Office of the Secretary of Defense, the Joint Staff, the Unified Combatant Commands, the Navy, the Marine Corps, the defense agencies, and the defense intelligence enterprise.

The research reported here was completed in February 2021 and underwent security review with the sponsor and the Defense Office of Prepublication and Security Review before public release.

For more information on the RAND Navy and Marine Forces Center, see www.rand.org/nsrd/nmf or contact the director (contact information is provided on the webpage).